S0-DLP-323

Dynamic Alliance Auctions

Contributions to Management Science

H. Dyckhoff/U. Finke
Cutting and Packing in Production and Distribution
1992. ISBN 3-7908-0630-7

R. Flavell (Ed.)
Modelling Reality and Personal Modelling
1993. ISBN 3-7908-0682-X

M. Hofmann/M. List (Eds.)
Psychoanalysis and Management
1994. ISBN 3-7908-0795-8

R.L. D'Ecclesia/S. A. Zenios (Eds.)
Operations Research Models in Quantitative Finance
1994. ISBN 3-7908-0803-2

M.S. Catalani/G.F. Clerico
Decision Making Structures
1996. ISBN 3-7908-0895-4

M. Bertocchi/E. Cavalli/S. Komlósi (Eds.)
Modelling Techniques for Financial Markets and Bank Management
1996. ISBN 3-7908-0928-4

H. Herbst
Business Rule-Oriented Conceptual Modeling
1997. ISBN 3-7908-1004-5

C. Zopounidis (Ed.)
New Operational Approaches for Financial Modelling
1997. ISBN 3-7908-1043-6

K. Zwerina
Discrete Choice Experiments in Marketing
1997. ISBN 3-7908-1045-2

G. Marseguerra
Corporate Financial Decisions and Market Value
1998. ISBN 3-7908-1047-9

WHU Koblenz – Otto Beisheim Graduate School of Management (Ed.)
Structure and Dynamics of the German Mittelstand
1999. ISBN 3-7908-1165-3

A. Scholl
Balancing and Sequencing of Assembly Lines
1999. ISBN 3-7908-1180-7

E. Canestrelli (Ed.)
Current Topics in Quantitative Finance
1999. ISBN 3-7908-1231-5

W. Bühler/H. Hax/R. Schmidt (Eds.)
Empirical Research on the German Capital Market
1999. ISBN 3-7908-1193-9

M. Bonilla/T. Casasus/R. Sala (Eds.)
Financial Modelling
2000. ISBN 3-7908-2282-X

S. Sulzmaier
Consumer-Oriented Business Design
2001. ISBN 3-7908-1366-4

C. Zopounidis (Ed.)
New Trends in Banking Management
2002. ISBN 3-7908-1488-1

U. Dorndorf
Project Scheduling with Time Windows
2002. ISBN 3-7908-1516-0

Tobias Ihde

Dynamic Alliance Auctions

A Mechanism for Internet-Based Transportation Markets

With 21 Figures and 31 Tables

Physica-Verlag
A Springer-Verlag Company

Series Editors
Werner A. Müller
Martina Bihn

Author
Dr. Tobias Ihde
DaimlerChrysler AG
Abt. RIC/EN
Alt-Moabit 96 A
10557 Berlin

ISSN 1431-1941
ISBN 3-7908-0098-8 Physica-Verlag Heidelberg New York

Cataloging-in-Publication Data applied for
A catalog record for this book is available from the Library of Congress.
Bibliographic information published by Die Deutsche Bibliothek
Die Deutsche Bibliothek lists this publication in the Deutsche Nationalbibliografie; detailed bibliographic data is available in the Internet at <http://dnb.ddb.de>.

Physica-Verlag Heidelberg
a member of BertelsmannSpringer Science+Business Media GmbH

Printed in Germany

Softcover-Design: Erich Kirchner, Heidelberg

SPIN 10942079 88/3130-5 4 3 2 1 0 – Printed on acid-free and non-aging paper

Acknowledgements

There are people to thank. I am grateful to Prof. Dr. Ulrich Schmidt and Prof. Dr. Andreas Drexl. They gave me both support and guidance whenever necessary but also the opportunity to go my own way and to follow my own curiosity. The working atmosphere was open-minded and very pleasant to me. I want to express my gratitude to Dr. Tibor Neugebauer, with whom I conducted the auction experiment. His experience on auction experiments and his advice were valuable for me. Thank you for your engagement.

This thesis is part of DaimlerChrysler's research project Virtual Truck Enterprise (VTE). I am grateful to my colleagues at DaimlerChrysler's research department RIC/EN who provided an inspiring and exciting environment. In particular, I would like to give thanks to Dr. Klaus Schild and Stefan Bussmann for many controversial and fruitful disputes, which frequently resulted in new ideas. I also want to thank Birgit Burmeister and Regina Lasch for the splendid team work during our research project. Also, I want to thank Dr. Sunil Thakar for various discussions on what really matters in writing a PhD thesis. Thanks go also to Dr. Kurt Sundermeyer for his support and engagement.

I want to give thanks to all my friends for general interest in my work and general support, especially to Marc Noske, Rik Uhmeier, Axel Wulf and Reza Kazemi-Tabrisi.

I am particularly grateful to my family. My mother, my father and my brother have always supported me best, and so did my grandmother and grandfather. Special thanks go to my wife Kathrin. During the writing of my thesis, Kathrin has had the necessary persistence and has always been a fan on mine. Thank you, it has helped a lot.

Tobias Ihde

Contents

1 Introduction

1.1 Motivation

Today's transportation market is far from perfect. Market participants have admission only to a certain fraction of the market and price formation lacks transparency. However, this state is no law of nature since in principle, all market participants could manage to get together somehow and do business – but gaining new business partners beyond existing bonds usually takes too high efforts. After all, the items in transportation markets are usually traded under extreme time pressure: transportation orders have fixed pick-up and delivery deadlines and free transportation capacity cannot be used thereafter. Here, Internet-based freight markets promise help. Since the early 1990s many of such markets have come into existence so that a considerable number exists today. For Germany alone, over 50 different electronic transportation marketplaces are listed in a database hosted at the University at Bremen (2001). The mediation models of these marketplaces are quite diverse (cp. Ihde (2002), Schneider et al. (2002)). Electronic blackboards with or without fixed fees, open only to a certain part of the market, can be found as well as auction houses and real exchanges with transaction-based fees (Teleroute (2001), Benelog (2001*a*), Eulox (2001*a*)). Nevertheless, one important aspect has been ignored up until now: carriers usually have non-additive preferences with respect to combinations of transportation orders. The underlying reason is that capacity utilization of the respective vehicles can be maximized by forming round trips, which will generally bring down transportation costs significantly. In the simplest (and ideal) case, a carrier who is located in A has one order from A to a location B and one order from B to A that he can fulfil both sequentially with one and the same vehicle. If two such 'complementary' orders are auctioned off packagewize, both sides of the market will benefit – carriers, because they have a maximal capacity utilization granted and can make better prices without shrinking margin. Shippers benefit from these better prices. Auctioning off two complementary orders as a package is easy as long as these orders belong to the same shipper. The crux, however, is that in transportation spot markets they will generally stem from two different shippers. This has two consequences. First, two shippers with complementary orders do have to know about this and they do have to get into contact with each other. Second, the hammer price for an order package will be a *package* price and the involved shippers have to agree about their respective shares. In a highly dynamic business like the transportation business with little time for coordination, these are serious obstacles. However, both obstacles could be overcome. The first problem can be tackled with software routines that automatically identify complementary orders. Such a routine has been developed at DaimlerChrysler's research department. The second problem requires the design of a mechanism that allows for a packagewize placement of orders from two different shippers. Obviously, the mechanism involves the collection of orders, the aggregation of complementary orders to order packages, the placement of orders and order packages, and the division of package prices. This second issue is what this thesis is about.

1.2 Environment and Aim of the Thesis

This thesis is part of the DaimlerChrysler research project VTE[1], which aimed at providing DaimlerChrysler's Internet-based transportation marketplace Fleetboard(2002) with an unprecedent mediation service. During this project, a new auction-based mechanism was developed that allows for a packagewize placement of two 'complementary' transportation spot orders and that imposes no additional effort for shippers and carriers. This mechanism is called *Dynamic Alliance auction* and to some extent it has already been explored by Ihde & Schild (2002).

The main task of this thesis was

- to introduce Dynamic Alliance auctions as a simple and robust mechanism that makes a packagewize placement of two complementary spot orders possible,
- to give an economic motivation for the proposed design,
- to analyze Dynamic Alliance auctions with respect to bidding behavior and the performance of the auction format.

1.3 Contribution of the Thesis

These are the contributions of this thesis.

1. Introduction of Dynamic Alliance Auctions: The thesis introduces the new auction format by a distinct set of rules.

[1] Virtual Truck Enterprises.

2. Economic Motivation: The thesis gives an economic motivation for the proposed design with respect to order collection, order aggregation, placement, and the division of package prices. Dynamic Alliance Auctions are shown to be robust against collusion and bid sniping. The aggregation of orders is proved to represent a two-sided stable matching (from an ex-post view). Furthermore, the thesis shows that the division of package prices represents an asymmetric Nash bargaining solution. For the placement stage, some modifications for the usage of combinatorial auctions (instead of standard auctions) together with Dynamic Alliance auctions.
3. Formal Models: The thesis establishes a modified private value framework to further analyze bidding behavior. It models two interesting special cases and develops a formal model for shippers' expected payoff in Dynamic Alliance auctions.
4. Analysis: Within the above private value model, the thesis analyzes bidding strategies and the subsequent outcome of Dynamic Alliance auctions in the two special cases and in general markets. For the special cases, Nash equilibria are identified. These equilibria are shown to break down in arbitrary markets. For the arbitrary markets intuitive ad-hoc strategies are delivered.
5. Experimental Exploration: Dynamic Alliance are experimentally explored with respect to bidding strategies and the subsequent outcomes and different informational treatments. One of these treatments is shown to induce the best performance of Dynamic Alliance auctions.

Almost as a side product, the thesis delivers a classification scheme for Internet-based transportation marketplaces.

1.4 Organization of the Thesis

After the Introduction, which forms the first chapter, the thesis is divided in three parts.

The first part is called **Background** and consists of two chapters. Chapter 2, named *Theories*, considers axiomatic bargaining, auctions, and the marriage problem. It starts with the introduction of the axiomatic approach to bargaining problems and then describes the famous solutions by Nash (1950) and Kalai & Smorodinsky (1975). Afterwards, it explores different frameworks in which auctions can be analyzed: the model of private values, the common value model and the model of affiliated values. A premium is put on the private value model. The last section of this chapter is on the marriage problem, which has been included because order aggregation in Dynamic Alliance auctions represents a two-sided matching problem.

Chapter 3 alias *Internet-based Transportation Marketplaces* portrays the fast-evolving landscape of Internet-based transportation marketplaces. For this purpose, it first delivers a classification scheme for such marketplaces (cp. Ihde (2002)), which will later be used to identify the marketplaces where Dynamic Alliance auctions can be applied. Second, the chapter explores three Internet-based freight marketplaces alongside the classification scheme, a trait to get across how differently such marketplaces work. These three marketplaces have been chosen because they represent the whole range of possible and actually operating market models.

The second part, which motivates the needs for a new mediation mechanism and which introduces Dynamic Alliance Auctions, is titled **Freight Auctions**. It comprises Chapter 4 and 5. Chapter 4 is called *Conventional Freight Auctions* and identifies the markets in focus of this thesis – Internet-based transportation spot markets. It explains the shortcoming of conventional freight auctions and shows why combinatorial freight auctions cannot be applied to transportation spot markets. Afterwards, it identifies requirements to be met by a mediation mechanism that is designed for placing packages of two complementary orders stemming from two different shippers. Chapter 5 – *Dynamic Alliance Auctions* – establishes Dynamic Alliance Auctions. It describes the stages through which this new auction form runs and gives a distinct set of rules for Dynamic Alliance auctions. The chapter concludes with an example that illustrates how Dynamic Alliance auctions work.

The third part is named **Evaluation** and includes Chapter 6, 7, 8, and 9. Chapter 6, called *Stages and Price Division*, aims at giving the economic motivation for the proposed design of Dynamic Alliance auctions. The mechanism is shown to satisfy the requirements formulated in the second part. Furthermore, the stages of Dynamic Alliance auctions and the rule for dividing package prices are linked to the context of auctions, the marriage problem and axiomatic bargaining. The chapter shows that the aggregation of orders leads to an two-sided ex-post stable matching and shows that the division of package prices represents an asymmetric Nash solution. Chapter 7, which is called *Efficiency, Payoff, and Bids*, establishes a modified private value framework and and analyzes the outcome of Dynamic Alliance auctions within this framework. It starts with the investigation of special cases and identifies Nash equilibria for them. Afterwards, it models a general payoff function for Dynamic Alliance auctions and shows that the special case equilibria break down in arbitrary markets. This breakdown is demonstrated in an exemplary market, which is experimentally explored in Chapter 8, called *Experiment*. This chapter describes the experimental treatments and reports on interesting findings. The thesis concludes with *Putting Insights to Practice*, which is Chapter 9. This chapter briefly summarizes the basic assumptions and the findings of this thesis. It also reflects practical implications and the needs for future research.

Part I

Background

2

Theories

2.1 Axiomatic Models of Bargaining

Bargaining refers to a situation in which two or more persons (*players*) are confronted with a set of feasible outcomes of which one will be realized if and only if an unanimous agreement on this outcome is reached. In case that no unanimous agreement is reached, a *conflict* or *disagreement outcome* will be the result. For the purposes of this thesis it is sufficient to consider bargaining models with two players. A simple example for such a bargaining situation is how to divide a dollar between two players. Suppose that two players are offered a dollar in case they can reach an agreement how to divide it. If they are not able to reach such an agreement, no player gets anything. This problem can be found quite often in reality, for example, if cost savings are possible but can only be gained by means of mutual cooperation. This section will introduce various solutions to bargaining games that satisfy 'reasonable' conditions – *axioms*. The axioms represent desirable properties a bargaining outcome should possess (Sandholm (1999)). Accordingly, the solutions are called *axiomatic bargaining solutions.*

2.1.1 The Basic Model

Formally, a 2-person bargaining game $(P, \tilde{c})$ consists of

- a compact convex subset P of $\mathbb{R}^2$,
- a disagreement outcome $\tilde{c} \in P$, which represents the outcome if no agreement is reached,
- two players, of whom each has a preference order on the set of feasible outcomes,
- von-Neumann-Morgenstern utility functions mapping the set of feasible outcomes onto P, which represent the players' preference orders.

Let $\mathcal{B}$ be the set of all bargaining problems. A *solution* to a bargaining problem is a function $f : \mathcal{B} \to \mathbb{R}^2$ such that $f(P, c)$ is an element of P for all bargaining games $(P, c) \in \mathcal{B}$ (Roth (1979); Owen (1995)).

2.1.2 Literature on 2-Person Bargaining

Axiomatic bargaining originated with the article by Nash (1950). Nash proposed the first axiomatic solution – the *Nash solution*, and should be regarded as obligatory reading. The same holds for the paper by Kalai & Smorodinsky (1975) who introduced an important alternative to Nash's solution. Roth (1979) provides a detailed technical study of axiomatic bargaining, and his website (Roth (2002)) represents a very comprehensive source of information on this and many other topics. Osborne & Rubinstein (1990) and Binmore & Dasgupta (1987) provide a thorough analysis of different models and connect bargaining to the study of equilibrium markets.

2.1.3 Solution Concepts

This section introduces several solution concepts – the Nash solution, the Kalai-Smorodinsky solution, the proportional solution, and the asymmetric Nash solution. Most of the content has been taken from Osborne & Rubinstein (1990).

The Nash Solution

The first axiomatic solution was proposed by Nash (1950). He showed that for certain axioms a unique solution exists – the so-called *Nash solution.* These axioms were the axioms of *invariance to equivalent utility transformations*, *symmetry*, *independence of irrelevant alternatives*, and *pareto-optimality.* They will be introduced and briefly described.

INV *(Invariance to equivalent utility transformations)* A bargaining solution f is said to be invariant to equivalent utility transformations if $f(P, c) = f(Q, d)$ holds for all bargaining problems (P, c) and (Q, d) for which $a \in \mathbb{R}^2_{>0}$ and $b \in \mathbb{R}^2$ exist such that $(Q, d) = (a \cdot P + b, a \cdot c + b)$.

SYM *(Symmetry)* A bargaining solution f is called symmetric if $f_1(P, c) = f_2(P, c)$ holds for all bargaining problems (P, c) such that $c_1 = c_2$ and for all $(x, y) \in P$ also $(y, x) \in P$.

IIA *(Independence of irrelevant alternatives)* A bargaining solution f is said to be independent of irrelevant alternatives if $f(P, c) = f(\tilde{P}, c)$ holds for all bargaining problems (P, c) and $(\tilde{P}, c)$ such that $P \supseteq \tilde{P}$.

PAR *(Pareto optimality)* A bargaining solution f is called pareto optimal or efficient if $f(P, c) \geq x$ holds for all bargaining problems (P, c) and all $x \in P$.

Axiom INV imposes that the players' preferences are decisive, and not the utility functions that represent them. If a player's preferences are represented by a utility function u, all utility functions $v := a \cdot u + b$ will represent the same preferences if $a > 0$ and b is a real number. Axiom INV 'requires that utility outcome of bargaining co-vary with the representation' (Osborne & Rubinstein (1990), p. 13), but the (arbitrary) choice of utility functions will not change the underlying solution outcome.

Axiom SYM rules out any differences in the players' bargaining abilities or bargaining power into account. Eventual asymmetries between the players should be modelled into (P, c).

Axiom IIA implies that only the disagreement vector and the solution outcome itself are relevant, and that the outcome does not depend on other alternatives in the set. In particular, all solutions that maximize the value of some function fulfill IIA. On the other hand, if the outcome represents a compromise between the (eventually incompatible) demands of players, IIA will not be satisfied. Consequently, and differently from the other axioms, IIA models the bargaining process itself. The appropriateness of IIA depends on the particular bargaining problem.

PAR requires that players behave 'collectively rational'. The solution will never lead to an outcome which is less preferred by both players than another feasible outcome. In this way PAR implies a solution's proofness against re-negotiation.

Nash (1950) showed that there is a unique bargaining solution that satisfies INV, SYM, IIA, and PAR. This function is called *the (symmetric) Nash solution* and will be denoted by f^N. To any bargaining game $(P, \tilde{c})$, f^N assigns

$$f^N((P, \tilde{c})) = \text{argmax } (x - c) * (x' - c') \quad \text{subject to} \quad (x, x') \in \mathcal{P}.$$

Of course, Nash's solution also satisfies the axiom of individual rationality.

IR (*Individual Rationality*) A bargaining solution f is called individually rational if $f(P, c) \geq c$ holds for all bargaining problems (P, c).

If any of the four axioms INV, SYM, IIA, or PAR is dropped, a solution can be found that satisfies the other three axioms but is not the Nash solution. Hence none of Nash's four axioms is superfluous (Osborne & Rubinstein (1990)).

The Kalai-Smorodinsky Solution

Kalai & Smorodinsky (1975) introduced an alternative solution that is based on the *ideal point* of a bargaining game. Formally, for any $(P, c) \in \mathcal{B}$, the ideal point $m(P, c)$ is defined by

$$m(P, c) := (m_1(P, c), m_2(P, c)),$$

where

$$m_1(P, c) := \max\{s_1 | (s_1, s_2) \in P\}$$

and

$$m_2(P, c) = \max\{s_2 | (s_1, s_2) \in P\}.$$

The ideal point consists of the maximum utilities that each player they could achieve in P and will generally not belong to P (see Figure 2.1).

The solution that Kalai & Smorodinsky (1975) introduced is called *Kalai-Smorodisky solution* and will be denoted by f^{KS}. For any bargaining game (P, c), the Kalai-Smorodinsky solution $f^{KS}(P, c)$ is the maximal point i in P on the line that connects c and m. Figure 2.1 gives an illustration.

The Kalai-Smorodinsky solution is the only solution that satisfies INV, SYM, PAR and the following axiom of *individual monotonicity*:[1]

IMON (*Individual Monotonicity*) A bargaining solution f is called individually monotonous if $f_j(P, c) \geq f_j(Q, c)$, for all bargaining problems (P, c) and (Q, c) and any $i \in \{1, 2\}$ such that $P \supseteq Q$, $m_i(P, c) = m_i(Q, c)$, and $j \neq i$.

The Kalai-Smorodinsky Solution and Roth's Ideal Point

An alternative ideal point $m'(P, c)$ can be found in Roth (1979). Roth's ideal point is made up by the maximal individually rational utilities for each of the players (instead of all feasible utilities). Formally, it is defined by

$$m'(P, c) := (m'_1(P, c), m'_2(P, c))$$

such that

$$m'_1(P, c) := \max\{s_1 | (s_1, s_2) \in P, s_1 \geq c_1, s_2 \geq c_2\}$$

and

[1] Note that Kalai & Smorodinsky (1975) simply refer to IMON as 'axiom of monotonicity'.

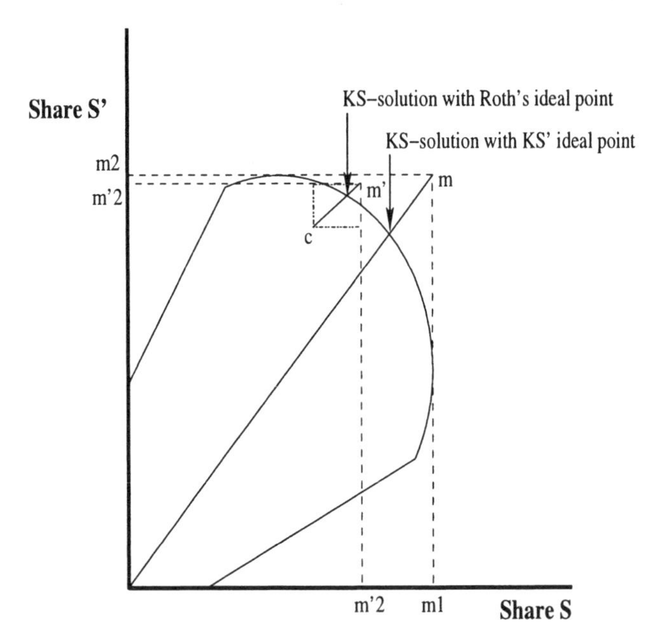

Figure 2.1. The Kalai-Smorodinsky solution with the ideal point of Kalai-Smorodinski (m) and with the ideal point of Roth (m'). (Figure is similar to Holler & Illing (1996), p. 208.)

$$m_2'(P, c) = \max\{s_2 | (s_1, s_2) \in P, s_1 \geq c_1, s_2 \geq c_2\}.$$

Roth's ideal point will lead to the same outcome for games in which the ideal point and Roth's ideal point are identical, but to different outcomes otherwise. Figure 2.1 depicts a bargaining problem where the two different ideal point m and m' lead to different solutions. The Kalai-Smorodinsky solution with Roth's ideal point is given by the point i', which is the interception of the boundary of P and the line connecting c and m'.

Solutions Satisfying SYM and PAR

All bargaining solutions satisfying SYM and PAR will lead to the same outcome as the Nash solution f^N for any symmetric bargaining game (Osborne & Rubinstein (1990)). Consequently, the Kalai-Smorodinsky solution will coincide with the Nash solution for symmetric games, regardless how the ideal point is defined.

Assuming linear utilities for both players, the game how to divide a dollar between two players is a symmetric game. It can be formalized as $(\tilde{P}, \tilde{0})$, where $\tilde{P}$ is the convex hull of the points $(0,0)$, $(1,0)$, and $(0,1)$ and $\tilde{0} = (0,0)$. The Nash solution is given by $f^N((\tilde{P}, \tilde{0})) = f^{KS}((\tilde{P}, \tilde{0})) = (1/2, 1/2)$.

The Asymmetric Nash Solution

An objection to Nash's solution concept is that is does not take differences in the players' bargaining abilities or bargaining power into account. Different bargaining abilities are likely to lead to a nonsymmetric outcome, even if the underlying bargaining set is symmetric (Holler & Illing (1996)).

A way to account for such asymmetries is to introduce a 'weight' $w \in (0,1)$ and to define the solution f^w by

$$f^w((P,\tilde{c})) = \text{argmax } (x-c)^w * (x'-c')^{1-w} \quad \text{subject to} \quad (x,x') \in P$$

for any bargaining game $(P,\tilde{c})$ and any $w \in (0,1)$.

The set $(f^w)_{w\in(0,1)}$ is called the family of *asymmetric Nash solutions*. For any w, f^w satisfies INV, IIA, and PAR (Binmore (1987)). Roth (1979) showed that also the other direction holds: any solution that satisfies INV, IIA and PAR equals f^w for some $w \in (0,1)$.

Applying the asymmetric Nash solution to the dollar example $(\tilde{P},\tilde{0})$ leads to $f^w(\tilde{P},\tilde{0}) = (w, 1-w)$ (cp. Osborne & Rubinstein (1990), p. 21).

The Proportional Solution

The basic idea of proportional solutions is to distribute gains in a bargaining situation in a fixed proportion between the two players. It is thoroughly investigated in Roth (1979).

The *proportional solution* is a function f^p that assigns to any bargaining game (P,c) and a vector $\bar{w} = (w, 1-w) > 0$ the outcome

$$f^p((P,c)) = \max Z \cdot \bar{w} + c \quad \text{subject to} \quad (Z \cdot \bar{w} + c) \in \mathcal{P}, \quad Z \in \mathcal{R}$$

The proportional solution satisfies IR, IIA, and IMON. It is the only solution that satisfies MON and DECO.

DECO (*Decomposability*) A bargaining solution f is called decomposable iff $f(P,c) = f(P, f(Q,d))$ whenever $(P, f(Q,d))$ is a bargaining game.
MON (*Monotonicity*) A bargaining solution f is called monotonous iff $f(P,c) \geq f(Q,c)$ for all bargaining problems (P,c) and (Q,c) such that $P \supseteq \tilde{P}$.

Axiom DECO allows a stepwise bargaining process. The players can bargain over a subset Q of P, such that Q contains c, and bargain over the entire set P afterwards, taking the result of the first bargain as conflict outcome. As illustration, Figure 2.2 depicts a bargaining solution that does not satisfy DECO: a stepwise bargaining leads to the outcome $i2$ whereas the 'one-step' approach results in i. Axiom MON ensures that all players benefit from any enlargement of the bargaining set. MON is a stronger axiom than IMON and is not necessarily compatible with PAR.

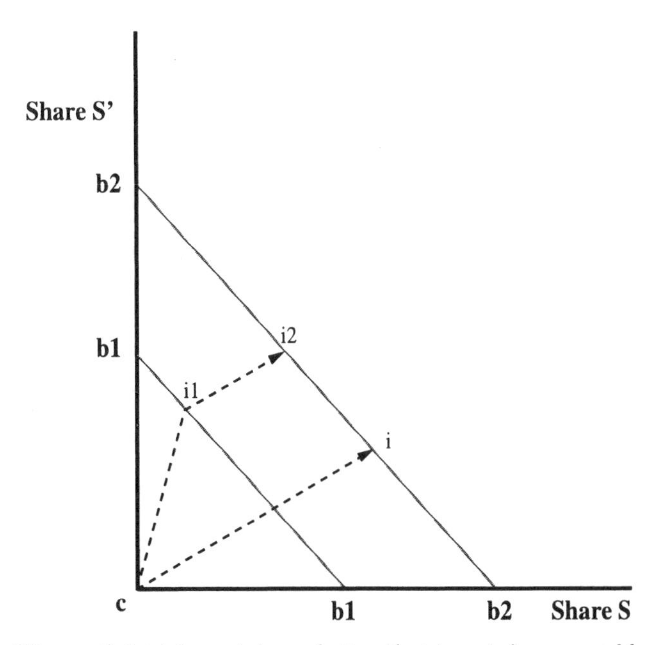

Figure 2.2. A bargaining solution that is not decomposable.

Applying the proportional solution with distribution vector $(w, 1-w)$ to the dollar example $(\tilde{P}, 0)$ leads to $f^p(\tilde{P}, \tilde{0}) = (w, 1-w)$. Here, $f^{p(w)}$ equals the asymmetric Nash solution f^w with weight w.

More generally, $f^{p(w)}$ equals f^N and f^{KS} on the class of symmetric games in case the distribution vector is $w = (1/2, 1/2)$. It is identical to an asymmetric Nash solution f^w with weight w.

2.1.4 A Note on the Strategic Approach to Bargaining

The strategic approach models the bargaining process as a sequential game. This section shows how the Nash solution is related to the solution by Rubinstein (1982) for strategic bargaining models. Most of it has been summarized from Owen (1995).

Re-consider the dollar example and assume the two players have linear utilities for money. In a strategic model, the two players will alternately make offers and counteroffers until an agreement is reached, i.e. until an offer or a counteroffer is accepted. Player 1 is supposed to make the first offer. He selects a point (u^*, v^*) from the boundary of the set $\tilde{P}$ (which is defined as in the previous sections) and offers v^* to player 2. Now player 2 has the option to accept or refuse. In case he refuses, the game enters the second stage, in which player 2 chooses a point (u^{**}, v^{**}) from the boundary of $\tilde{P}$ and offers u^{**} to player 1. Now it is player 1's turn to accept or refuse the offer. If he refuses, the game enters the third stage, where player 1 can make an offer again. In principle, infinitely many stages can be played. To take time consumption into account, *discount factors* are introduced into the model. Each player i $(i = 1, 2)$ is assumed to have a discount factor $\delta_i < 1$.

Rubinstein (1982) showed that there is a uniquely determined point (u^*, v^*) on the boundary of $\tilde{P}$ such that

- player 1 should always offer v^* and accept any offer of at least u^*,
- player 2 should always offer u^* and accept any offer of at least v^*,
- no player can benefit from acting differently in one stage if the other player always sticks to the recommended strategy.[2]

The pareto-optimal point (u^*, v^*) is called the *Rubinstein solution* and given by $u^* := (1-\delta_2)/(1-\delta_1\delta_2)$ and $v^* = 1-u^* = \delta_2(1-\delta_1)/(1-\delta_1\delta_2)$. The Rubinstein solution (u^*, v^*) actually represents what the two players in the dollar example will get in a strategic bargaining model. In case both discount factors are equal and approach 1, the according Rubinstein solution approaches the outcome of the Nash solution, which was $f^N(\tilde{P}, \tilde{0}) = (1/2, 1/2)$.

2.2 Auctions

The previous section dealt with one-to-one bargaining situations. However, in many situations a person faces not only one but many potential partners with whom he could do business. For instance, a seller of a unique painting could face N potential buyers. He might ask himself which price he should demand.

In case the seller knows buyers' valuations everything is fine. He sets the price equal to or slightly below the highest valuation and extracts all the surplus. On the other hand, if there was a 'standard value' for unique paintings, he could use that value as a posted price. Generally of course, the seller will neither be informed that well nor will there be a standard price. Without auctions several one-to-one negotiations, interleaved and interdependent, will often have to be conducted. This procedure is cumbersome and time-consuming,

[2] Put simply, Rubinstein showed that a unique subgame perfect equilibrium in stationary strategies exist.

and it will generally not lead to an efficient outcome. More complications arise if the owner of the painting employs an agent to sell the painting. Hence it will be a much better idea to conduct auctions in most cases.[3] Auctions can be considered as highly formalized 'direct extensions of the usual forms of bilateral trading' (Wilson (1992), p. 228).

Summarizing, the main reasons for the widespread use auctions are simplicity, speed of sale, and ignorance of what price to post represent. Preventing selling agents from dishonesty is important as well (Wolfstetter (1999); Rothkopf & Harstad (1994)).

2.2.1 Terminology

Auctions are formal trading mechanisms that determine price and allocation of one or more items for sale. They are based on an explicit set of rules (McAfee & McMillan (1987)). These rules specifies the procedure of bidding, the winning bid and the price at which an object is awarded to or *knocked down to* one of the bidders – the so-called *hammer price*. Also, there will generally be a *reserve price*, set by the seller, and if the bids do not exceed this price the item will go unsold and be called *bought in* (Ashenfelter (1989), p. 24).

Depending on the rule for bidding, auctions can be divided into *open* and *closed* auctions. In open auctions, all bidders get immediately informed if a new bid has taken the lead, and at which price. In closed auctions bids are disclosed to the auctioneer only (Rothkopf & Harstad (1994)). In fact, there is no standard terminology on this issue. Open auctions are also called oral, open cry-out auctions or open-bid auctions. On the other hand, closed auctions are also called written or sealed-bid auctions or closed-bid (Wolfstetter (1999); Wilson (1992); McAfee & McMillan (1987)).

There are four auctions that are regularly referred to as the *standard auctions* (Matthews (1995); Milgrom (1989); Wolfstetter (1999)): the English auction, the Dutch auction, the first-price auction, and the Vickrey auction. The English and the Dutch auction represent open auctions in contrast to First-price and Vickrey auctions, which are closed (see Table 2.1).

Open	Closed
English (Ascending-price)	Vickrey (Second-price)
Dutch (Descending-price)	First-price

Table 2.1. 'The most popular auctions.' (Taken from Wolfstetter (1999), p. 184.)

[3] This does not just hold with respect to transaction costs but also judging by the expected price (cp. Bulow & Klemperer (1996)).

The English Auction

The English auction (also called *Ascending auction*) is an open auction and most frequently used. A seller starts the bidding by announcing a minimum price and asks for a bid higher than this price. If such a bid occurs, other bidders are allowed to overbid it. Every bid is disclosed to all other bidders. The bidding procedure continues until an ultimate bid is made, i.e. a bid that is not overbid anymore. The winning bidder pays a price equal to his bid. For ease of exposition, the following variant is often considered (cp. Wolfstetter (1999)): the auctioneer starts with an minimum price which is displayed on an 'English clock'. Bidders may stay active of leave the auction while the displayed price is continuously being raised. Once a bidder has left the auction, he is not allowed to get into it again. The auction continues until exactly one active bidder is left. This bidder gets the item and is charged the price displayed on the clock when the last but one bidder quits. No other bidder is charged a payment.

The Dutch Auction

The Dutch auction (also called *Descending auction*) is an open auction and owes its name to its use for selling cut flowers in the Netherlands. An initially high price is continuously being lowered until the first bidder accepts. The current price is the price this bidder has to pay (cp. Milgrom (1989)).

The First-Price Auction

The first-price auction belongs to the class of closed auctions. Each bidder submits just one sealed bid, without knowing the bids of his rivals. The highest bid wins and the winning bidder has to pay the price of his bid (cp. Milgrom (1989)).

The Vickrey Auction

The Vickrey auction also represents a closed auction. Bidding procedure and winner determination are identical to the first-price auction. The only difference consists in the pricing rule: the winner has to pay the second highest bid instead of his own (the highest) bid (Vickrey (1961)). Consequently, Vickrey auctions is also called second-price sealed-bid auctions.

Using Auctions

While Vickrey auctions are rarely met in reality, the first three auctions are widespread and frequently used. In fact, auctions have been used for thousands of years for a bewildering variety of items: for Babylonian slaves, the Roman Empire (Shubik (1983); Cassady (1967)), from art to wine (cp. Ashenfelter (1989)) but also for transportation orders (cp. Ihde (2002)). Historical sketches are given by Shubik (1983) and Cassady (1967), while Lucking-Reiley (2000) surveys recent auctions on the Internet. Note that in industrial procurement, auctions are generally conducted as so-called *reverse* auctions. In a reverse auction, there is one buyer and several sellers. Most results hold for them also.

2.2.2 Literature on Auctions

The seminal paper of Vickrey (1961) marks the beginning of the academic analysis of auctions. Since Vickrey's paper, a large body of literature on auction theory has developed. Since it is highly dispersed the literature guide by Klemperer (2002) is extremely useful. Detailed technical primers have been contributed by Wolfstetter (1996, 1999) and Matthews (1995). They represent thorough introductions into the topic. Wilson (1992) gives also a technical analysis of auctions. Classical surveys over the theory of auctions are those by McAfee & McMillan (1987), Milgrom (1989; 1987; 1985) and Wilson (1987). A survey over the field of experimental research of auctions can be found in Kagel (1995). Additionally, the critical essay by Rothkopf & Harstad (1994) should be obligatory reading because it draws the necessary attention to existing gaps between the theory and the practice of auctions.

2.2.3 Solution Concepts

There are many different aspects of auctions to analyze. For a seller, the natural starting point is how to maximize his revenue and how to prevent collusion. Economists generally ask how efficient an outcome will be. Bidders are usually interesting in the 'best' *bidding strategy*, i.e. a function that takes a bidder's valuation as input and has the price of his bid as output.

To look at the bidding strategies is the key to most questions because any outcome depends on the bids submitted. Consequently, the common approach to 'solve' auctions is to find states from which no bidder wishes to deviate – so-called *equilibria* – and analyze their outcome. The following equilibrium concepts are used in this thesis.

Strong Equilibria

A strategy for a bidder is called *(strongly) dominant* if it maximizes his utility regardless which strategies his rivals chose. If all bidders pursue strongly dominant strategies, the vector comprising all strategies is called a *strong equilibrium*. It represents the strongest solution concept for auctions. Unfortunately, the concept of strong equilibria is often too strong. In general, bidders will have to take their rivals' bids into account.

Nash Equilibria

Given the strategies of the competing bidders, a bidder's choice of a bidding strategy can also be considered as his choice of a probability to win the auction and the according payment – i.e. as his choice of a lottery. Suppose there are N bidders and they have chosen the strategies $b_1^*, \dots b_N^*$. Let $u_i(b_1^*, \dots, b_N^*, v_i)$ denote the utility of bidder i associated with his lottery. For risk-neutral bidders, u_i can simply be regarded as the expected payoff.

The strategy b_i^* is called *weakly dominant strategy* for bidder i if

$$u_i(b_1^*, \dots, b_i^*, \dots, b_N^*, v_i) \geq u_i(b_1^*, \dots, b_{i-1}^*, b, b_{i+1}^*, \dots, b_N^*, v_i) \tag{2.1}$$

holds for any b and any value v.

Under the assumption that all other players stick to their strategies, bidder i cannot gain from changing his strategy. Obviously, the situation is stable if equation (2.1) holds additionally for any bidder. This is the concept of a Nash equilibrium:

The N-tuple of strategies $(b_1^*, \dots b_N^*)$ is called a *Nash equilibrium* if equation (2.1) holds for holds for any b and any value v and any $i = 1, \dots, N$. A Nash equilibrium $(b_1^*, \dots b_N^*)$ is *symmetric* if additionally $b_1^* = b_2^* = \dots = b_N^*$ is given. Nash equilibria provide a common solution concept to auctions.

2.2.4 Theoretical Models

To make further analysis of bidding behavior possible, some assumptions have to be made. Now auction models come into play. In the literature, three basic models are discussed – the model of *independent private values*, the model of the *common value*, and the model of *affiliated values*. In each model, there is generally a fixed number N of bidders considered who join the auction.

The Model of Private Values

The model of symmetric independent private values (SIPV) is the best known auction model. In this model, the following prerequisites are supposed to hold (cp. Matthews (1995); Wolfstetter (1999)):

A1 *Indivisibility:* One single indivisible object is put up for sale to one of the N bidders.
A2 *Symmetry:* Values are identically distributed random variables.
A3 *Private Values:* Each bidder knows his value, but he does not know other bidders' values.
A4 *Independence:* Values are independent.

A5 *Risk Neutrality:* All bidders and the seller are risk neutral.

A1 is clear. A2 (the assumption of symmetry) abstracts from different bidder 'classes' like e.g. foreign companies and domestic companies, dealers and collectors or, a subtle but relevant distinction, more sophisticated and less sophisticated dealers. A3 is called the private value assumption. Except the private values of other bidders, everything else is commonly known, including the distribution of values in A3. Due to A3, it is not possible for any bidder to draw conclusions from his value on values of other bidders. A4 also rules out any possibility of reselling the item after the auction has ended. If a bidder had a resale in mind before or during the auction, he would have to take into account the price he could later charge. But at this moment, the price will generally be unknown so that information on valuations of eventual buyers would be valuable information. In the course of the auction, signals about other bidders' valuations could serve our bidder as such information and make him adapt his own valuation to these signals, in contradiction to A4. Risk neutrality (A5) implies that seller and bidders maximize their expected profit. As stated before, bidders play lotteries by choosing a bid. A risk neutral person is someone who will play the lottery with the higher expected value if offered the choice between two lotteries. Thus in an auction, a risk neutral bidder will submit the bid with the highest expected profit. In the context of risk neutrality, dominant strategies and Nash equilibria refer to the expected payoff.

The Common Value Model

In the common value model the item for sale has the same 'true value' for every bidder – this is the common value. However, no bidder knows this value exactly (Milgrom (1989)). Oil rights represent an example for a common values situation. The principally extractable amount of oil will be the same for all bidders. But bidders have different information about this amount and are not certain. This is why Myerson (1981) interprets a common value situation as a situation of quality uncertainty.[4] In such situations, learning other bidders' information (valuation) is crucial. In the light of new information a bidder can update his own valuation and change his bidding behavior.

An important insight for common value auctions is the *winners curse.* To have won a closed auction for an item with a common value is, in fact, bad news. The winner was more optimistic than his competitors and therefore submitted a higher bid. An instructive example is the experiment of Bazerman & Samuelson (1983). In this experiment, Bazerman & Samuelson filled jars with coins and auctioned them off to students, using first-price sealed bid

[4] Formally, bidders' valuations are independent draws from some probability distribution, conditioned on the common value. All bidders know this distribution.

auctions. The value of the jars – Dollar 8 each – was kept a secret before and during the auctions. After all the jars had been sold, it turned out that the average winning bid was Dollar 10.01 while the average bid was only Dollar 5.13. Experiencing the winner's curse, the average winner made a loss. To put this insight to sealed-bid auction practice, each bidder should assume that the own quality signal was the highest and revise the own assessment downwards (McAfee & McMillan (1987)).

The Model of Affiliated Values

The model of affiliated values was introduced by Milgrom & Weber (1982) and embraces both the common value and the private value model. The principle of *affiliation* means that for bidders with a high quality signal it is likely that the other bidders have received similarly high quality signals. Formally, the valuation of any bidder is modelled as a function of all bidders' quality signals and the seller's quality signal. However, since the quality signals and valuations of other bidders are not known to the considered bidder in general, he does not know precisely how much the item is worth for him.

Milgrom & Weber (1982) show that, in equilibrium, the seller expects the highest revenue in an English auction, the second highest in a Vickrey auction, and the third highest profit in a Dutch auction. The expected revenue of a first-price auction equals the expected revenue of a Dutch auction. When facing affiliated values, the English auction generates the highest expected revenue because

> '[...] the bids in the English auction have the effect of partially making public each bidders' private information about the item's true value, thus lessening the effect of the winner's curse.' (McAfee & McMillan (1987), p. 722.)

In Vickrey auctions, the seller can increase his expected payoff by always making his private signal public. This insight is called *linkage principle.* However, the linkage principle depends heavily on the prerequisites and is hardly to be generalized. The linkage principle does not hold in multi-unit auctions, if symmetry is given up or if the assumption of common values is softened (cp. Wolfstetter (1998)).

2.2.5 Revenue, Efficiency, and Collusion in Private Value Auctions

This paragraph deals with a seller's revenue in standard auctions, the efficiency of auction outcomes, and the possibility of collusion. Most content of this paragraph has been compiled from Milgrom (1989), Wolfstetter (1999), and Matthews (1995).

Revenue

One fundamental result in auction theory is the *revenue equivalence theorem.* It is due to Vickrey (1961).

Theorem 2.1. *(Vickrey (1961)) In equilibrium, the seller expects the same revenue in English, Vickrey, First-price and Dutch auctions.*

The return equivalence theorem follows from three observations:

1. English and Vickrey auctions yield the same expected equilibrium revenue,
2. Dutch and First-price auctions yield the same expected equilibrium revenue,
3. the expected revenues in 1) and 2) are the same.

In English auctions, it is a dominant strategy for bidders to stay in the auction until the demanded price exceeds their valuation. Hence the equilibrium hammer price will be the second highest valuation. In a Vickrey auction, the hammer price does not depend on the highest but on the second highest bid. Hence bidding below one's valuation reduces the probability of winning while it does not affect the price in case of indeed having won. Conversely, bidding above one's valuation may lead to a loss if winning and changes nothing if not. Consequently, deviation from truthtelling does not pay and, again, the hammer price equals the second highest value (cp. Milgrom (1989), p. 8). Putting this together, English and Vickrey auction yield the same expected revenue for the seller.

In first-price and Dutch auctions, bidders must privately determine a certain price and offer to buy the auctioned item for this price. In both auctions, the bidder with the highest offer is awarded the item and pays the price of his bid. Since the rules for allocation and pricing are identical, first-price and Dutch auction are strategically equivalent (Milgrom (1989)). In these auctions, a bidder's profit-maximizing bid depends on his rivals' bidding strategies, so that the concept of weakly dominant strategies must be adopted and a Nash equilibrium is searched for.

Both first-price and Dutch auctions have a unique symmetric equilibrium of bidding strategies where 'each bidder's equilibrium bid is equal to the expectation of the maximum of his competitors' values conditional on that value being less than his own'. (Matthews (1995), p.18.). Formally, with N bidders whose values $v_1, \ldots, v_N$ are independently identically distributed according to an $F : [0, a] \to [0, 1]$, the equilibrium bidding strategy is given by

$$B(v) := v - \int_0^v \left(\frac{F(x)}{F(v)} \right)^{n-1} dx \tag{2.2}$$

(compare McAfee & McMillan (1987), p. 709). Denoting the second order statistic of the values by $v_{(2)}$, the right-hand side of(2.2) equals

$$E[v_{(2)}|v_{(2)<v}], \tag{2.3}$$

which is the formal expression of Matthews' (1995) statement.[5] The seller's expected revenue in this equilibrium equals the expectation of bidders' second highest valuation. This is the same expected revenue as in English and Vickrey auctions.

In fact, equivalence of revenue applies not only to the standard auctions but to all auctions with a symmetric equilibrium in which the bidder with the highest valuation is awarded the item (Myerson (1981); Riley & Samuelson (1981)). However, it depends heavily on the assumptions (A1-A5). If symmetry is given up (A2), revenue equivalence result does not hold anymore (Vickrey (1961); Maskin & G. (2000); Wolfstetter (1999)). It breaks also down if bidders are not risk-neutral (A5) (McAfee & McMillan (1987); Klemperer (1999)) or if the independence assumption (A4) is dropped. If the private-value assumption (A3) is softened in the sense that bidders do not have not full certainty about their valuation but can invest money in order to obtain it, counterspeculation against other bidders' strategies may pay (Sandholm (1999)). Consequently, it is now longer clear that the equivalence of revenue still holds.

Revenue equivalence in private value auctions has also been explored experimentally. According to Kagel (1995), the finding is that revenue equivalence does not seem to hold, but risk attitudes of bidders have generally not been controlled for. English and Vickrey auctions did not lead to comparable outcomes: while subjects quickly transit to truthtelling in English auctions, they generally bid above their values in Vickrey auctions. In both Dutch and first-price auctions bidder seem to bid above the equilibrium bids. As a rule, auctions where bidders must specify a bid (first-price and Vickrey auctions) lead to higher prices than open auctions where prices are announced an bidders simply must accept or reject (English and Dutch auctions) (cp. Kagel (1995)).

Efficiency

An auction is called efficient if it leads to a Pareto-optimal outcome, i.e. if the bidder with the highest value wins the item. Under the assumption that the seller always sells the item, all standard auctions are efficient (Wolfstetter (1999)).

If a reservation price exists, efficiency may not necessarily hold in Vickrey auctions (Milgrom (1989)). In case the highest bidder bids more than

[5] Proofs can be found in Wolfstetter (1999, p. 195 ff, and 1996) and Matthews (1995).

the reserve price but the runner-up bids less, the outcome is inefficient. If the symmetry assumption is knocked off the English and the Vickrey auction stay efficient, provided that the seller always sells the item. On the other hand, this is not true for Dutch and first-price sealed-bid auctions (Wolfstetter (1999)).

In experiments, efficiency of private value auctions can be measured as the percentage of auctions with Pareto-optimal outcome. Kagel (1995) reports on experiments where efficiency was 79% in Vickrey auctions, between 82% and 88% in first-price auctions and about 80% in Dutch auctions.

Collusion

An important practical issue concerning auctions is to prevent of collusion. Presuming private values, McAfee & McMillan (1992) showed that a potential bidding ring can find a truthful bidding inducing mechanism to designate a winner and calculate the side payments. This makes the question whether some auction types are more susceptible by collusive agreements than others even more important. Robinson (1985) showed that such agreements are self-sustaining in an English auction but not in a Dutch auction. The argument is as follows. In an English, the designated winner could always bid up to his own valuation while all others bid zero. No member of the bidding ring has an incentive to play differently. In a Dutch auction, however, the designated winner must bid slightly more than the reserve price, while all other ring members bid zero. Of course, ring members can generally gain from bidding slightly higher than the designated winner (Klemperer (2002)).

Collusion in common experimental settings is generally not reported on (Kagel (1995)). Isaac & Walker (1985) conducted an experiment to investigate collusion in sealed-bid auctions. Although no side payments were allowed, they found that stable collusion occurred in the majority of auctions. Typically, the bidder with the highest value won the item with a minimal bid while the other bidders bid zero.

2.2.6 Combinatorial Auctions

Many auctions serve for the sale of several distinct items instead of just one item. Due to complementarities or substitutional effects, bidders often have non-additive preferences for combinations of items. *Combinatorial auctions* represent an auction form that allows bidders to express these preferences by bidding on combinations instead of the corresponding single items. At present, combinatorial auctions are being tested at DaimlerChrysler (Ihde et al. (to appear)), and they have already been used by the American company Sears Logistics Services (Ledyard et al. (to appear)). Sears Logistics Services auctioned off 3-year-contracts for 854 lanes. The auction was conducted in five bidding rounds. In each round, 14 pre-qualified carriers were free to bid for

each of the $2^{854} - 1$ combinations theoretically possible. As a result, Sears Logistics Services were able to bring down their transportation costs \$190000000 to \$160000000, i.e. by approximately 13%.
A survey on this topic has been contributed by DeVries & Vohra (2000), which forms the basis for this section together with the paper by van Hoesel & Müller (forthcoming).

In their purest form, combinatorial auctions work as follows for super-additive preferences:

(1) Each bidder is allowed to submit a bid for any combination.
(2) Each item can be sold only once.
(3) Those combinations are selected that
  a) do not violate (2) and
  b) lead to the maximal sum of bids.

Combinatorial auctions require the solution of an optimization problem. Suppose there are N bidders and K items. Let J_i denote the number of bids submitted by bidder i, $i = 1, \ldots, N$. Define $a_{ijk} := 1$ if the j-th bid of bidder i is on a combination that contains item k and 0 otherwise. Then the optimization problem can be formalized as

$$\max \sum_{i=1}^{N} \sum_{j=1}^{J_i} b_{ij} x_{ij} \tag{2.4}$$

$$s.t. \sum_{i=1}^{N} \sum_{j=1}^{J_i} a_{ijk} x_{ij} \leq 1 \qquad \forall k = 1, \ldots, K \tag{2.5}$$

$$x_{ij} \in \{0, 1\} \qquad \forall i = 1, \ldots, N, \forall j = 1, \ldots, J_i. \tag{2.6}$$

Unfortunately, this optimization problem is hard to solve.[6] However, if it can be solved (on time), efficiency will maximal. Consequently, the auctioneer faces a trade-off between efficiency gains and the complexity imposed on bidders (and computers). In the case of the FCC spectrum rights in the United States in 1994, it was decided not to use combinatorial auctions:

> 'In the judgement of most of the economists involved in the auction design, the complexity costs outweighed the potential efficiency gains: the full-combinatorial mechanism was ahead of its time.' (McMillan (1994))

Nevertheless, combinatorial auctions are discussed as a significant method to reduce transportation costs.[7] In transportation business, they are used to place transportation orders and, of course, conducted as *reverse* auctions. Accordingly,

[6] It belongs to the class of $\mathcal{NP}$-complete optimization problems.
[7] Further examples for the application of combinatorial auctions can be found in e.g. Brewer (1999), Caplice (1996) and Rassenti et al. (1982).

(1) Each bidder is allowed to submit a bid for any combination of orders.
(2) Each order can be placed only once, and all orders must be placed.
(3) Those order combinations are selected that
   a) do not violate (2) and
   b) lead to the minimal sum of bids.

The corresponding optimization problem can be formalized as

$$\min \sum_{i=1}^{N} \sum_{j=1}^{J_i} b_{ij} x_{ij} \tag{2.7}$$

$$s.t. \sum_{i=1}^{N} \sum_{j=1}^{J_i} a_{ijk} x_{ij} = 1 \qquad \forall k4 = 1, \ldots, K \tag{2.8}$$

$$x_{ij} \in \{0, 1\} \qquad \forall i = 1, \ldots, N, \forall j = 1, \ldots, J_i. \tag{2.9}$$

In this model, equation (2.8) formalizes condition (2).

2.3 The Marriage Problem

The so-called marriage problem represents a two-sided matching problem, where members of one group must be matched to members of another group. In fact, the marriage problem occurs quiet often in reality, for instance on the job market, where employees and companies must be matched. Two-sided matching has been scrutinized by Gale & Shapley (1962), Roth (1984; 1990), and Mongell & Roth (1991)). Roth's website contains valuable links, bibliographies and short introductory article to this topic Roth (2002). Significant questions of matching problems involve stability of matchings, (i.e. proofness against re-negotiation), matching procedures and manipulation. Most of the content of this section has been borrowed from Wolfstetter (1999).

2.3.1 Stability

Formally, a marriage problem is made up by a group of *women*, which is a set $\mathcal{W} := \{W_1, \ldots, W_n\}$, and a group of *men*, which is a set $\mathcal{M} := \{M_1, \ldots, M_n\}$. It is assumed that

- $\mathcal{W}$ and $\mathcal{M}$ are disjoint,
- each man has complete, transitive, and strict preference order on $\mathcal{W}$,
- each woman has complete, transitive, and strict preference order on $\mathcal{M}$.

The aim is to build couples (man, woman) that are stable in the sense that no man and no woman can be found who are not married with each other but who both prefer each other to his/her present spouse. Definition 2.2 formalizes stability.

Definition 2.2. *A matching is called* stable *if for all pairs* (W, M_j) *and* (W_k, M) *the following holds:* M *prefers* W_k *to* W *or* W *prefers* M_j *to* M. *A matching that is not stable is called* unstable.

It turns out that for all preferences, the marriage problem has at least one stable matching. Two simple procedures can be used to obtain it.

2.3.2 Matching Procedures or Who-Proposes-to-Whom

The first procedure is the *'man-proposes-to-woman'* procedure and ends after at most n^2 steps. Each step comprises two actions. In the first step,

- Each man proposes to the woman he prefers most.
- Of all man that proposed to her, each woman accepts the man she prefers most as her suitor but rejects all the others.

In every step that follows,

- Each man rejected in the previous step proposes to the woman he prefers most.
- Of all man that proposed to her, each woman accepts the man she prefers most as her suitor but rejects all the others.

The second procedure works identically except for fact that men and women change their roles – now women propose to men. Logically, it is named *'woman-proposes-to-man'*. However, the role change is significant since both procedures may produce different stable outcomes as the following example will show. Take the case of three men M_1, M_2, and M_3 and three women W_1, W_2, and W_3. Suppose that the preferences are as in Table 2.2. Each cell shows how a man ranks a woman (first number) and this woman ranks this man (second number). For instance, M_1 prefers W_1 the most while he is only the second-best choice for W_1. Table 2.2 shows that each man prefers

	W_1	W_2	W_3
M_1	1,2	2,3	3,1
M_2	4,3	1,1	2,2
M_3	3,1	4,2	1,3

Table 2.2. Preferences of men and women.

the woman with the same index the most. Consequently, the man-proposes-to-woman procedure will match each man with his most preferred woman. The corresponding couples are $(M_1, W_1), (M_2, W_2)$, and (M_3, W_3). On the other hand, each woman favors the man whose index adds with hers to 4, and the woman-proposes-to-man procedure will give rise to the couples $(M_1, W_3), (M_2, W_2)$, and (M_3, W_1) (cp. Wolfstetter (1999)).

2.3.3 Reporting Preferences

As the example indicates, each procedure will generally be in favor of one of the two groups. In case there is more than one stable matching, the man-proposes-to-woman procedure selects the outcome that is collectively favored by men.[8] It is a dominant strategy for every man to reveal his true preference. However, this is not true for women, who can gain from strategic reporting under the man-proposes-to-woman procedure. On the other hand, the outcome collectively preferred by women is selected if women propose to men. This time, revelation of true preferences is dominant for woman but not for man. Putting this all together no procedure can be found that leads to stable outcomes and that makes reporting the true preferences a dominant equilibrium for all preferences. This statement remains valid if replacing the term 'strong equilibrium' by 'Nash equilibrium'.

Despite strategic reporting, the two procedures will produce stable outcomes (Roth & Sotomayor (1990)). If the best strategic preference reporting is chosen by women under the man-proposes-to-woman procedure, the matching is stable with respect to both reported and true preferences. The same holds for the matching outcome if women propose to men.

[8] A proof can be found in Wolfstetter (1999), p. 178.

3

Internet-Based Freight Marketplaces

3.1 A Classification Scheme

Basically, a marketplace can be defined as a location where demand for a good meets supply for this good so that trades can be made. A marketplace is Internet-based if the Internet is used as communication medium. With this principal definition three issues are sufficient to characterize Internet-based marketplaces – they make up their core.

- **Participants**: Who are the participants of the marketplace?
  This issue specifies who hosts the marketplace and who has admission to it. It also determines the market structure and the number of industries that the trading agents belong to.
- **Traded Goods**: What is the object of trade?
  This question covers all relevant aspects related to the traded good. The good traded in transportation marketplaces is transportation capacity. Transportation capacity is no homogenous good. Transportation mode, terms, regional restrictions and vehicle requirements represent its key constitutional points.
- **Trade**: How will trade be coordinated and facilitated?[1]
  The market mechanism fixes how trade is coordinated and whether it takes place on the marketplace itself or not. Hence which market mechanism is deployed plays a substantial role. The questions who initiates trade and to which extent business transactions are supported are also relevant: they have strong implications for the fee model. Facilitation of (buyer) cooperation may represent a new and interesting criterion as well.

The answers to the above questions form the building blocks for the marketplace. Changing one of them is changing the marketplace – which is why

[1] Technological aspects of Internet-based marketplaces are beyond the scope of this thesis. However, these aspects are covered by the issue 'trade' since technological issues always aim at enhanced usability, i.e. facilitation of trade.

these questions represent a good guideline for the following exploration of Internet-based freight markets. A similar guideline along which Internet-based procurement markets are investigated is used in Segev et al. (1999) and Merz (2002).

In an Internet-based transportation market, transportation capacity represents the traded good, and demand and supply for this capacity are matched via Internet. The efforts to establish transportation marketplaces based on information and communication technology date back to the early Seventies (cp. Büllingen (1994)). Nevertheless, only few categorizations for Internet-based transportation marketplaces are available (Bretzke et al. (2001), Schneider et al. (2000) and Polzin (1998)). The systematic rigor and the foci of the respective studies vary substantially and with them the extent to which certain aspects are covered. Schneider et al. (2000) surveys existing Internet-based transportation marketplaces and discusses actual trends and obstructions for these marketplaces along eight characteristics all of which can be assigned to the three issues given above.[2] The approach taken is a practical one. On the other hand, Bretzke et al. (2001) approach transportation markets from both a theoretical side, in which transportation marketplaces are classified[3], and a practical side. The latter contains the results of interviews with users of transportation marketplaces and a survey of European and American transportation marketplaces with a description of the respective services. Polzin (1998) investigates logistical services and their suitability for electronic trade. For this purpose, a thorough and useful characterization of logistical services and markets is given.

The following establishes a grid for Internet-based transportation marketplaces. Together with the case study, this grid conveys the great variety of possible and already existing transportation marketplaces.

3.1.1 Participants

The **host** of a marketplace can either be a buyer of transportation capacity, a seller of transportation capacity or a neutral third party offering the information technological infrastructure. Who runs the marketplace represents a relevant criterion since neutrality is generally considered as key success factor in the literature. The reasoning is that neutral hosts is supposed to optimize the marketplace with respect to the needs of as many carriers and shippers as possible and not just with respect to his own needs (cp. Polzin (1998)). Merkel & Kromer (2001) see a lack of neutrality result in competitors refusing the marketplace.

---

[2] These characteristics are branch of industry, number of trading agents, regional restrictions, fee model, technological requirements and quality management.

[3] However, the classification is rather coarse. The host, structure, number of industries and order terms are the relevant criteria for Bretzke et al. (2001).

With regard to **admission**, transportation marketplaces can either be public or private. If agents willing to participate in trade have admission provided they pass certain quality thresholds, the marketplace is *public*. It is *private* if only selected agents do have admission. Admission rules are relevant since they must fit the actual market situation. They also influence the number of buyers and sellers in the marketplace.

The number of buyers and sellers of transportation capacity determines the **structure of the market**. The structure may span the whole spectrum from bilateral monopoly (one seller facing one buyer)[4] over monopoly and monopsony to polypoly (many sellers facing many buyers).The market structure has important implications for the trading mechanism as Section 3.1.3 will show.

The **number of industries** that buyers and sellers stem from is widely used as criterion for classifying Internet-based marketplaces. A transportation marketplace supporting transactions of only one industry is called *horizontal* marketplace. As a rule, horizontal marketplaces offer highly detailed functionalities that ease trade within a certain industry, e.g. the chemical industry. In contrast to horizontal marketplaces, *vertical* marketplaces offer a smaller portfolio of services for all kinds of industries. Hence this criterion could also have been placed under the issue traded goods.

3.1.2 Traded Goods

Loads can be moved via road, air, sea or railway. This is referred to as **transportation mode**. Most Internet-based transportation marketplaces focus on road haulage; only buyers and sellers of *truck* capacities are matched on such marketplaces. Shippers buy transportation capacity for cargo by placing transportation orders and carriers sell transportation capacity by accepting those orders. Forwarding agents fulfill transportation orders themselves but also subcontract orders to carriers, so they can act both as shippers and carriers.

Orders mediated in a transportation market can either be short-term (*spot market*) or long-term (*contract market*). In a spot market, short-term demand and supply are matched. Spot orders are usually of no strategic importance. On the other hand as a rule contracts are strategically relevant, and often the specificity is high. Contracts regulate frame conditions for the fulfillment of concrete transportation orders in contract markets (cp. Bretzke et al. (2001), p. 4). Hence personal negotiation is necessary so much so that standard auctions are inappropriate, not to mention double auctions. With respect to significance, long-term contracts outweigh short term orders by far: according to

[4] Bilateral monopoly is not the most desirable scenario for the host.

Bretzke et al. (2001), 86% of shippers' transportation orders represent long-term contracts, and 70% of carriers' capacity are bound to long-term agreements. Consequently, the **terms** of mediated transportation orders represent a vital criterion for classifying transportation markets because it may limit the possible degree of automation in mediation. Hence not all mechanisms presented in Section 3.1.3 may be appropriate.

All transportation orders specify a pick-up and a drop location for the respective load. Internet-based transportation marketplaces usually restrict themselves to certain **regions**. These regions can be whole Europe, a part of Europe or only one country (cp. Teleroute (2001), Benelog (2001*a*), LSXS (2001)).

The **vehicle requirements** of orders vary greatly. Basically, it is possible to differentiate between *standard* loads and *non-standard* loads and between *full* loads and *partial* loads (the latter referred to a certain truck size). Standard loads can be transported with standard trucks using standard pallets. On the other hand, non-standard loads need special trucks or special treatment. The more standardized an order is the better it is suited for electronic trade.[5] Moreover, vehicle requirements represent one of the keys to determine which market mechanism is best suited, as the discussion in the following section and Figure 3.1 will show.

3.1.3 Trade

Today's Internet-based transportation marketplaces usually deploy blackboards or auctions as **market mechanisms**. Sometimes also double auctions can be found. The choice of an appropriate mechanism, i.e. a mechanism that fits the market situation, is crucial as the following will show.

An electronic *blackboard* is a database that can be accessed through the Internet. Shippers can use it to post their orders. Carriers search it and, if an appropriate order is found, the corresponding shipper is contacted via phone, fax or email. One-to-one negotiations follow and finally a carrier is awarded the freight contract. This procedure can also be vice versa: carriers post their free capacity, shippers search the blackboard and negotiations follow. Electronic blackboards can also contain take-it-or-leave-it-orders, orders supplemented with a non-negotiable price. In traditional transportation business, these offers are commonly used by shippers for consigning commodities since here bargaining costs usually outweigh bargaining savings.

In an *auction* for transportation capacity one carrier auctions off free transportation capacity between several shippers, and the highest bidding shipper

[5] Closely related is how specifiable an order is. The easier an order can be specified, the more it is suited for electronic trade (cp. Polzin (1998), Merz (2002)).

wins. In a *reverse auction* there is one shipper who auctions off a job or contract between several carriers and the carrier with the *lowest* bid wins.

In a *double auction* for transportation capacity, carriers specify their vehicles and offer the corresponding capacities on routes of their choice for a certain price whereas shippers post orders together with a maximum price. Compatibility of requirements presumed, the highest bidding shipper is matched with the lowest bidding carrier and a price is calculated.

So the **initiative** for trade can be taken by

1. shippers: a shipper offers a transportation order and one or more carriers submit bids for this order. The shipper selects one of the bids and makes a contract with the respective carrier.
2. carriers: a carrier offers available transportation capacity and one or more shippers submit bids for its usage. The carrier selects one of the bids and makes a contract with the respective shipper.
3. both shippers and carriers simultaneously. Both parties simultaneously submit buy and sell offers for transportation capacity. Supply and demand are matched by a mediator.

Mechanism (1) and mechanism (2) seem to be equivalent but in fact they are not because in each case the party confronted with the price pressure is different. Mechanism (1) favors shippers whereas (2) puts carriers in a more comfortable position. The underlying reason is that with mechanism (1), carriers have to compete for the lowest price, while it is the shippers' turn with mechanism (2). Both mechanisms are implemented by means of auctions or blackboards.

Mechanism (3) represents an exchange-like mechanism similar to that of stock or electricity markets. But when comparing these markets to freight markets, it becomes quickly clear that freight markets are of a peculiar nature. To fulfill a transportation order, a vehicle must be moved to a loading point, to one or more corresponding unloading points afterwards and finally back to its home base or to the loading point of the following trip. Consequently, the specific routes, vehicle requirements and time restrictions must be accounted for. Hence transportation capacity cannot easily be turned into a commodity like 'capacity per mile for a certain period'. Nevertheless attempts were made to establish Internet-based transportation marketplaces using exchange-like mechanisms.[6] These marketplaces tried to tackle variety by restricting themselves to standard loads like palletized freight or container freight.

In today's freight markets, mechanism (1) is predominant, a fact reflecting the existing oversupply of transportation capacity. Price pressure rests on

[6] These were Eulox, and, according to Schneider et al. (2000), Back-Pack.

the shoulders of the carriers. This is why usually shippers initiate trade, no matter whether blackboards or auction mechanisms are deployed. Transportation auctions are generally conducted as reverse auctions. This has influenced the use of language. The term *freight auctions* is generally used for reverse auctions. This use of language is adapted for the remainder of this thesis. Unless stated differently, freight auctions stand for shipper-initiated auctions and therefore represent reverse auctions.

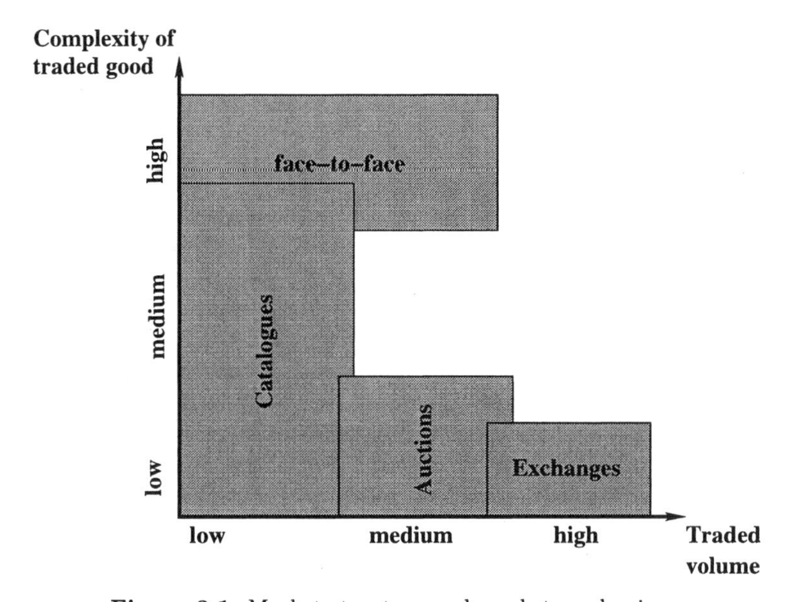

Figure 3.1. Market structure and market mechanisms.

Figure 3.1 depicts market mechanisms and their appropriateness for different market situations. Face to face negotiations cover highly complex goods. Catalogues are suited for products of low and medium complexity and a low trading volume. Auctions can do with (and need) a 'medium' trading volume and can handle products of preferably low complexity. Double auctions help to deal with (and need) high trading volume but are restricted to true commodities. Applied to transportation business this means that the use of double auctions is restricted to standard complete loads on highly frequented routes.

As stated at the beginning, a marketplace is a location where demand meets supply, where information is provided and where trade may happen. Trade means allocation of the good(s) and determination of the price. Both allocation and price determination create effort. From the viewpoint of market participants, not all mechanisms are equally 'convenient' since this effort is different for different market mechanisms. Figure 3.2 captures this idea for a many-to-many market. The horizontal axis depicts different market mechanisms, the vertical axis stands for allocation and price determination effort in a many-to-many market. The area above the bend line symbolizes the effort market participants have while the area underneath represents the effort reduction due to the mechanism. Interleaved one-to-one negotiations do not relieve the market participants from any of the burden, in contrast to double auctions, which reduce effort maximally. Catalogues and auctions lie somewhere in between.

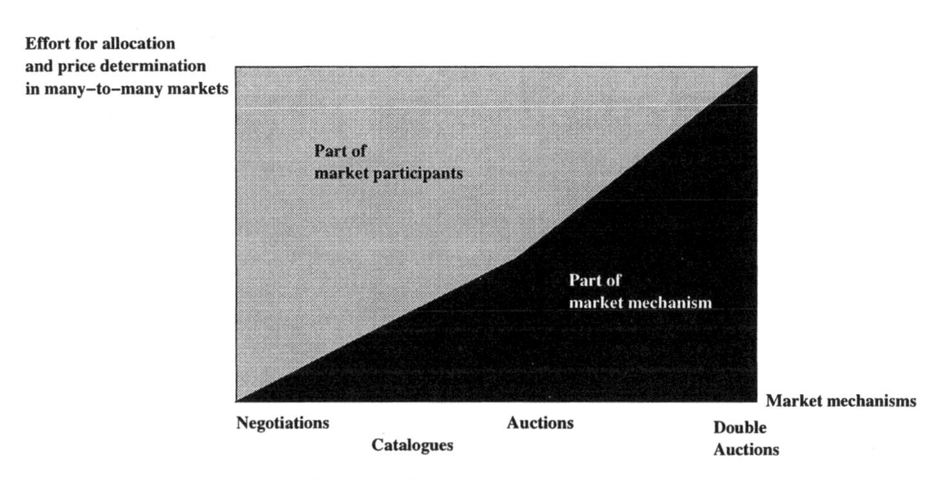

Figure 3.2. Effort for allocation and price determination.

The extent the marketplace host can support business transactions is related to the 'convenience' of a market mechanism. Put differently, the support of business transactions determines whether a marketplace is merely a point of aggregated information, which requires an additional communication medium to initiate and finish trade, or whether it is also the point of trade and transactions. The market mechanism deployed is the decisive aspect here. Business transactions run through three stages: the information stage, the agreement stage, and the stage of settlement (cp. e.g. Schmid (1993)).[7] Market participants gather information concerning services, products or partners and specify supply and demand in the *information stage*. An *agreement stage* follows where negotiations take place and, if successful, a binding contract is made. Contract execution makes up the last stage called *settlement stage* and in

[7] This stage model was later refined by Schmid (1998; 2000), but for the present purpose, the presented model will do.

transportation markets this stage includes for example assessment of transport operation or payment handling. **Online transactions** demand that all stages be processed online. The agreement stage is the crucial one. Since online contracting presupposes online pricing, the market mechanism plays the decisive role in the agreement stage. It determines allocation of items and relieves market participants from some of the burden of price negotiation. Some of the negotiation effort will always remain, but *how much* depends on the actual mechanism. The less market participants have to negotiate themselves, the easier online contracting can be supported. So the task is to chose and to implement the market mechanism that fits the market and reduces negotiation effort to a minimum. Consequently, two types – rather generations – of Internet-based transportation markets can be identified, those that do not provide online transactions and those that do. The issue of online transactions is closely related to the issue of the market mechanism deployed. In online auctions and double auctions, the first two stages are automatically processed online. On the other hand this is not necessarily the case in blackboards or catalogues.

How will the marketplace host generate profit, i.e., which kinds of fee will he **charge**, and whom? A great variety of different fee models exists, including all shades between basic fees, transaction-based fees and no fees at all. Principally, both shippers and carriers could be charged. But like the market mechanism, fee models usually mirror the distribution of market power. As a rule the weaker side of the market is charged, i.e. carriers. This resembles procurement markets (cp. Segev et al. (1999)). Closely related to the fee model is the *critical mass* of a marketplace – the number of traders necessary for the marketplace to survive. Bretzke (2001) roughly estimates that the critical mass for a marketplace mediating national freight orders in Germany is about 3,000 transportation orders per day. But such estimations must be treated with caution since the critical mass heavily depends on the business model and on the fee model in particular (cp. Merkel & Kromer (2001)). Consequently, there is no one critical mass that fits all marketplaces.

The fact that **cooperation** between shippers and between carriers can bring down transportation costs significantly is known in both the academic and the practical world (cp. Büllingen (1994)). Hence facilitation of cooperation seems to be an interesting but yet new criterion for classifying transportation marketplaces. Marketplaces might explicitly support and facilitate cooperation but generally do not.

3.2 A Critical Analysis

The peculiarities of the traditional transportation business must be accounted for when designing Internet-based transportation marketplaces. On the other hand, the possibilities offered by such marketplaces influence traditional transportation business back. Consequently, Internet-based transportation marketplaces have to be continuously adapted and improved. This all contributes to a very dynamic environment so much so that the following case study represents a snapshot of the perpetually changing landscape of Internet-based transportation marketplaces. Teleroute, Benelog and Eulox demonstrate how diverse Internet-based transportation marketplaces can be, which is why these three have been chosen for the following study. But since the case study does not only describe the status quo but also critically questions the approaches taken, it gets across principal ideas in the design of such marketplaces.

3.2.1 The Blackboard Teleroute

Characteristics

Participants

Teleroute is a horizontal marketplace founded in France in 1986. It is run by the publishing company Wolters-Kluwer, which represents a neutral third party. Only carriers and forwarding agents have admission to the blackboard, shippers are not allowed. Teleroute has managed to build up a customer base of 35,000 registered customers and with it the necessary trading volume.

Traded Goods

Teleroute mediates regional, national and international freight orders in Europe. All kinds of loads are permitted: partial as well as full loads and standard as well as special loads (Teleroute (2001)). Each day some 40,000 transportation orders are mediated by Teleroute, the lion share in France and approximately 8,000 odd orders in Germany. Short and long-term orders can be found.

Trade

Teleroute deploys a blackboard on which mainly forwarding agents initiate trade by posting orders. Carriers are charged a monthly basic fee combined with a fee for downloading order packages or read accesses to the database.

Analysis

Judging by Teleroute's trading volume, which is by far the largest of all European Internet-based transportation markets, the race to become *the* European Internet-based transportation market seems to be run since trading volume is a key factor for success in a network economy.[8] But Teleroute carries a historical burden that its competitors may capitalize on: shippers have no admission

[8] For a thorough treatment of network economies, trading volume and lock-in effects see Shapiro & Varian (1999).

to Teleroute's marketplace, a trait to preserve the traditional ecosystem of forwarding agents and carriers. Traditionally, shippers engage preferably forwarding agents for their transportation orders. Carriers hardly have access to shippers and get their orders subcontracted through forwarding agents. Consequently, carriers strongly depend on their forwarding agents. This is reflected in Teleroute's mediation model: transportation orders are not offered by shippers but by forwarding agents. Carriers must contact these forwarding agents to get load and so the traditional dependency is maintained. However, the general situation might change by means of Internet-based freight markets that offer online bidding. For shippers, placing a transportation order via electronic auctions or exchanges means no extra effort. On the contrary, it presents them with a possibility to save money. Due to an increased competition among carriers and a decreased number of mediation stages (*desintermediation*), prices are likely to decline so that electronic freight markets are quite attractive for shippers, and there is a good chance that they will use them. Following a KPMG study (Bretzke et al. (2001)), 94% of all interviewed shippers plan to place freight orders via electronic freight markets within the next three years. Since shippers generate transportation orders and not forwarding agents do, the number of orders placed by forwarding agents is likely to decline. Hence keeping away shippers from its marketplace might turn out to be disadvantageous for Teleroute.

Two more issues may be harmful for Teleroute: the fee model and insufficient actuality of freight data. How much a carrier must pay for a downloaded order should depend on whether he gets the order and what the order is worth. But this information is not available for Teleroute. Consequently, Teleroute can make its fee depend neither on the success of a mediation nor on transaction value. For carriers this is no good situation. A monthly basic fee means an expense with uncertain return and lock-in. Additionally, orders that have already been placed may still be on offer in the database because freight forwarders themselves have to remove them manually – after all, Teleroute does not know whether transactions occurred and, consequently, cannot do this. Hence data quality highly depends on freight forwarders' reliability. The fact that Teleroute has to empty its database every midnight shows that this reliability is not developed too well. From a carrier's perspective paying for a package of freight orders in advance is as if buying a pig in a poke. To pay for ten potential orders and then discover that three offers are antiquated is rather annoying – even more since also time and money have to be invested to contact the respective forwarding agents via phone or fax. KPMG's study (Bretzke et al. (2001)) indicates that fees based on transaction volumes seem to be preferred by the majority in the logistics sector. In case this can be generalized, there are probably many of Teleroute's customers who have an incentive for switching to a competing freight market.

The described difficulties both have the same root: lacking online transactions. Online transactions demand that all transaction stages (information, agreement, settlement) be processed online. Teleroute lacks online transactions because the agreement stage is not supported, i.e. no binding agreement can be reached on Teleroute's platform. Teleroute could overcome these problems through take-it-or-leave-it offers that carriers can accept via mouse click. For Teleroute, they would have two advantages. The first is accelerated formation of prices. The second is that these offers do not upset Teleroute's existing customers – in contrast to auctions, which regularly are suspected to spread more transparency than wanted (by Teleroute's customers).

However, the ostensible virtue of take-it-or-leave-it offers in Internet-based freight markets seems to be a half-way thought. No matter whether online freight auctions are conducted or take-it-or-leave-it-orders are deployed – a shipper's or forwarding agent's effort is the same. He must specify an order and set a price. In an auction this price is the *reservation price*, a maximal price which bids must not exceed. The auction itself follows clearly defined rules and does not require its initiator to supervise it. Thus using an auction instead of take-it-or-leave-it-prices can lead to savings for a shipper but not to additional costs or efforts.

3.2.2 The Auction House Benelog

Characteristics

Participants

Benelog is a horizontal marketplace and started its business in March 2000 and up until Mai 2001, Benelog had been able to register 2,200 companies as customers. The marketplace is open to shippers, carriers and forwarding agents.

Traded Goods

Benelog mediates single transportation orders, order packages and freight contracts for all kinds of truck loads in Europe. The majority of the contracts is short-term, but some long-term contracts are also mediated (*binding freight contracts*) and *tenders* without binding details of quantity. Each day about 100 orders are placed, so the critical mass has surely not been reached yet.

Trade

Reverse auctions play the dominant role for Benelog (2001*a*). The outcome of auctions is considered as binding contract.[9] A carrier winning an auction is

[9] In the U.S., agreements concluded online represent legally valid contracts. In Europe, corresponding laws have yet to be enacted.

charged a fixed fee for single national trips and single international trips. For binding freight contracts and auction tenders without binding details of quantity, the winning carrier pays a certain percentage of the transaction volume as commission. Since no other fees are charged, all fees are transaction-based and to be paid by carriers.

Registered shippers can chose between reverse *English* auctions (called *best-bid auctions*) and *price-quality auctions.* In a (reverse) English freight auction carriers must underbid each other with openly submitted bids. Carriers can respond to a bid by submitting a lower bid. Ultimately, the lowest bid wins and the winning carrier is paid the price of his bid. On the other hand, carriers do not have to underbid each other in price-quality auctions. In fact, their bids do not even have to be lower than the shipper's reservation price. The shipper can take quality aspects into account and select the mostly preferred bid.

Each Benelog auction has a predetermined duration of 30 minutes. If bids are received shortly before the auction ends, it is prolonged by three minutes. If again bids are received shortly before the prolonged auction is over, it is prolonged by another three minutes. This goes on until at most 60 minutes, the longest possible duration of Benelog's auctions. But the end of an auction does not depend solely on the incoming bids. As a specialty of Benelog's auctions, the initiating shipper is

> '... free to chose the end of an auction: he can let the process run automatically ("Best Bid") or intervene at any time and award the contract to a carrier.' Benelog (2001*b*)

Shippers may also define lists of carriers that may participate in the auction process and lists of carriers that must not. If an admissible carrier wants to enter an auction, he has to name the vehicle with which to fulfill the order in case of winning and to submit an 'initial bid' until at latest five minutes before the auction starts. His identity is not disclosed to anyone involved in the auction at this point.

Carriers can participate in several auctions at the same time ('parallel auctions') and are allowed to name the same vehicle, say 'X', in several parallel auctions. If an auction is won for X, all bids in parallel auctions for which X had also been named are removed.

After the auction, all formerly anonymized data is made known to both shipper and winning carrier. To exclude any transmission errors, the carrier has to confirm receipt of the freight data. After the transportation job has been finished, carrier and shipper have to assess each other with respect to punctuality, completeness, correctness of details etc. Carriers may also assess shippers' paying habits. For each shipper and carrier, all available assessments are aggregated to a ranking, which is intended to make quality differences transparent. Putting this together: information stage, agreement stage and settlement stage are processed online so that all transactions are online.

Analysis

Commercial auction websites like eBay had a roaring success and it seemed to be a self-evident step to transfer the auction mechanisms used to other Internet-based markets as well. But the success of standard auctions for selling single consumer goods does not automatically qualify them for other markets as well. For instance, many of Benelog's auctions are conducted simultaneously, so for carriers it is vital to bid in several parallel auctions with one and the same vehicle. Hence it makes sense that they are allowed to at Benelog's marketplace. Unfortunately, a problem occurs: if a carrier is awarded an order in one of the parallel auctions, his bids for this vehicle are removed from *all* of the remaining parallel auctions. This service makes sense while the orders in question exclude each other due to time or capacity restrictions, but it is annoying if this was not the case. The only possible work-around for carriers is to use different names for one and the same vehicle when bidding in parallel auctions for orders that do not exclude each other.

> 'Initially, Internet technology was used to try to build marketplaces that conformed to the simple model of a public auction. But real markets are much messier.' Varian (Dec 12th, 2000)

Benelog's decision to introduce price-quality auctions was probably devoted to the insight that transportation markets are far from being perfect. But the design of price-quality auctions is a delicate business.[10] In conventional auctions the bidding process is guided by one parameter: price. The lower the price, the better the bid. In a price-quality auction, at least two parameters determine which bid is best: price and quality. Attractive as Benelog's price-quality auctions may sound, they have a cumbersome feature, especially if many carriers participate – an unguided bidding process.

Suppose that carriers with different ranks participate in an auction. For the worst ranked carrier, the situation is clear. He must submit the lowest bid because with equal prices a shipper will always choose the better quality/ranking. But here clarity ends for him. Comparing to other bids, *how much lower* must he bid at least? He does not know. Analogously, the carrier with the highest rank may submit the highest bid, but he faces the same problem: how much higher at most? The situation is even worse for all carriers ranked between the best and the worst among participating rankings: they do not

[10] An interesting practice-driven approach has been contributed by Burmeister et al. (2002).

know at all in which price regions to bid. The underlying problem is that the carriers are given no information about how the shipper assesses trade-offs between bidden prices and existing ranks. Strictly speaking, the only way to overcome this problem is to give carriers this information after each bid – which requires the shipper's supervision or a utility function that reflects his preferences and is made known to all carriers. Both means disproportional effort in transportation spot markets. Hence it makes no sense to conduct *time*-based price-quality-auctions in a spot market since bidders do not know which bid leads and how to improve it. *Round*-based price-quality auctions may represent a better approach, among them, one-shot probably the best.

Business expertise, flair and some experimentation are also needed to design conventional price-driven auctions.Benelog's auctions are sometimes criticized for taking too much time. As Meier (2001) stated, a carrier will not have the time to bid 45 minutes for a single spot order if he has to acquire cargo for some dozens of vehicles.[11] Auctions with one or two bidding rounds could turn out as be a better solution again: they are fast and will not suffer from unwanted phenomenons like 'bid sniping'. Sniping is the name for a frequently observable behavior in Internet auctions with a predetermined ending time. In the course of an auction only few bids are submitted, but shortly before it ends, bidders 'jump in' and try to make the last bid. This behaviour appears regularly in eBay (2001) auctions which have fixed deadlines. Even sophisticated sniping programmes esnipe (2001) exist which allow to automate this behaviour. According to Roth & Ockenfels (2002) bidders will usually bid less than their true reservation price in auctions with sniping. Consequently the initiator of such an auction is worse off than in an auction without sniping. Since there are at most ten bidding prolongations for Benelog's auctions, they are not immune against sniping – at least theoretically. Practically, sniping has not turned out as a problem for Benelog, but one should keep it in mind if dealing with auctions that have a predetermined ending time.
On the other hand, it must be taken into consideration that transportation business is highly dynamic and that carriers might easily miss shorter or round-based auctions due to an urgent unforeseeable event. Setting the duration of electronic freight auctions is a tricky task. As already emphasized: it takes business expertise, flair and some experimentation to find a good compromise.

[11] Compare also Schneider et al. (2000)

3.2.3 The Exchange Eulox

Characteristics

Participants

Eulox went online at the beginning of 2001 and claimed to be a virtual forwarding agent, so that Eulox was not neutral but a seller of transportation capacity. Shippers, carriers and forwarding agents had access to the horizontal marketplace. Neither the number of registered users has been made publicly known and nor the number of transportation orders mediated per day.

Traded Goods

Eulox focus exclusively standard full loads between or in the regions Rhine – Ruhr, Paris – North France and East Austria (Eulox (2001*c*)). Additionally, swap and return of equipment – pallets and skeleton boxes – are offered. However, it is safe to suppose Eulox not to have reached its critical mass yet.

Trade

For each successfully mediated order a fixed fee is charged. This fee is paid in equal shares by the shipper and the carrier who have been matched. Eulox' market mechanism is extraordinary with regard to two aspects. First, a true exchange mechanism (a double auction) is deployed, which alone is somewhat revolutionary in transportation markets. Second, this exchange mechanism is a modified *double Vickrey auction.* A closer look shows that this is striking.

A common Vickrey auction is a *second-price sealed-bid auction.* The winning bidder (who is a buyer) has to pay the price of the second highest bid. In a *reverse* Vickrey auction the bidder (who is a seller) with the lowest bid wins the order and is paid the price of the second lowest bid.

In a double auction both sellers *and* buyers submit bids. Since, generally, the buyers' second highest bid will differ from the sellers' second lowest bid, the extension of a Vickrey auction to a double auction raises the issue of how the final price should be calculated. The price could be the second highest bid submitted by the sellers, the second lowest bid submitted by the buyers or some number in between. The fee for mediation must also be included.

Consider a specific route. As described in Section 3.1.3, registered carriers specify their vehicles, offer the respective capacities for this route and set a reservation price. A carrier's reservation price represents the minimum payment he wants to receive for an order. On the other hand, registered shippers offer loads for consignment. They too set reservation prices, each representing the maximum price a shipper is willing to pay for getting his load transported. As a specialty of Eulox, bidders do not stipulate just one reservation price but a whole range of reservation prices Eulox (2001*b*). After a carrier has submitted a bid, the maximum of his range is taken as reservation price. As time goes by, this is gradually being lowered until the minimum of the range is reached. For shippers, analogously, the minimum is taken at the beginning and gradually being increased until the maximum is reached. For ease of exposition and since at any time the reservation price of a chosen bidder is unique it is henceforth spoken simply of *the* reservation price.[12] Provided that the second highest shipper bid exceeds the lowest carrier bid by at least Eulox' fee, Eulox matches the lowest bidding carrier with the highest bidding shipper. The price the shipper has to pay and the payment the carrier gets do not depend on their own reservation prices. Instead, they are based on the mean value of the second highest shipper bid and the second lowest carrier bid as follows. Suppose that the two lowest carrier bids are 1,000 Euro and 1,100 Euro and that the two highest shipper bids are 1,250 Euro and 1,200 Euro. Presently, the Eulox fee for a successful mediation is 38 Euro so that involved shipper and carrier have to pay 19 Euro each. Then the shipper bidding 1,250 Euro is matched with the carrier demanding 1,000 Euro. The shipper pays $(1,200 + 1,100)/2 + 19 = 1,169$ Euro and the carrier gets $(1,200 + 1,100)/2 - 19 = 1,131$ Euro.

If a matching occurs, the involved carrier is informed that a transportation order is available. He then has five minutes in which he has the exclusive right to decide whether he accepts or rejects the order. In contrast to the carrier, the involved shipper is bound to his offer. Carriers are guaranteed to receive the freight rates within the agreed on time. On the other hand, Shippers are guaranteed to get their load transported.

Analysis

Eulox' matchings represent options to carriers but commitments to shippers. These issues present interesting asymmetries in the exchange mechanism and seem to be necessary adoptions to the requirements of transportation business.

According to Eulox, their mechanism leads to prices that are

> 'better for the shipper and haulier than those in a normal auction' Eulox (2001*b*)

and establishes a 'fair-price-system' Eulox (2001*b*). This provokes questions how prices that are *better for both sides* can constitute, what *fair* prices may be and whether these slogans are merely marketing slogans.

The statement that prices are 'better for both sides' must clearly be rejected. Since Eulox explicitly refers to 'normal auctions' in contrast to their

[12] The analysis to come also holds when taking into account the dynamic adjustment of reservation prices.

double auction the proclaimed 'better prices' must result from the specific type of auction Eulox deploys. But it is clearly impossible that both shippers and carriers get better prices simultaneously as they face a constant-sum-game in which Eulox' fee is the constant. To make this point clear: A shipper pays what his carrier gets plus 38 Euro for Eulox. Since Eulox' fee is the same for all transactions – 38 Euro –, a better price for the shipper means a worse price for the carrier and vice versa. So there is no way making both parties better off simultaneously by simply changing the auction type. Better prices for both sellers and buyers at the same time are possible only if the number of intermediation stages decreases (*disintermediation*) or if transaction costs sink. This may well be the case with Eulox – but also with other Internet-based freight markets using 'normal' auctions, for instance Benelog. Moreover, it is by no means clear which side in fact benefits from Eulox' exchange mechanism since in a common (one-sided) reverse Vickrey auction the expected final prices do not have to be higher than in a reverse first-price auction. This conclusion may look quite intuitive because the winning carrier is paid more than he has demanded, but it is also wrong since a change of auction rules results in a change of bidding behavior. Depending on the situation, the expected prices can be higher or lower than in a first-price auction.[13] A *double* Vickrey auction seems to be far more complicated.

Now for fairness. Vickreys auction induce *truthtelling*, i.e. in a common reverse Vickrey auction it is a dominant strategy for every carrier to bid exactly his true costs (cp. Chapter 2). As a consequence, Vickrey auctions are efficient – the carrier with the lowest costs wins a reverse Vickrey auction and in this sense the resulting price can be regarded as fair. Indeed this seems to be what Eulox refers to. However, some objections against this argument have to be made. First, it is by no means safe to transfer results from one sided auctions to exchanges. Second, a major problem concerning Vickrey's upshot is that it is based on special prerequisites that do not hold in general (cp. Wolfstetter (1999)). Without these prerequisites, the transportation domain is an El Dorado for counterexamples (cp. e.g. Sandholm (1999)). Third, Eulox' bidders do not set just one reservation price but reservation price *ranges*. Even if interpreting the adaption of reservation prices as time based discounting, it remains questionable if the price ranges are compatible with the fairness concept described above – why should a whole continuity of prices be fair? Putting this together, there seems to be no reason why Eulox' mechanism should lead to fair prices in any sense.

But Eulox' mechanism is also problematic for some other reasons. Suppose that two carriers, $C1$ and $C2$, demand 1,010 Euro and 1,015 Euro, respectively, whereas two shippers, $S1$ and $S2$, offer 1,100 Euro and 990 Euro, respectively. Principally, a trade between $S1$ and $C1$ is possible for some price

[13] Compare Milgrom & Weber (1982) and Wolfstetter (1999), chapter 8.

between 1,010 Euro and 1100 Euro. But since $S2$ bids less than $C1$ demands, no trade will be made.

This example shows that Eulox' mechanism does not necessarily exploit the whole potential for trading, in other words, their mechanism is not pareto-efficient. Generally, this inefficiency will occur whenever $p_{S1} > p_{C1} + f > p_{S2}$, where f is Eulox' fee, and p_{S1}, p_{S2} and p_{C2} are the bids of $S1$, $S2$ and $C2$, respectively. So the problem is that the second highest shipper bids less must be higher than the highest carrier bid (plus the fee).Inefficiency comes also up if carriers face only one shipper or vice versa. Strictly speaking, no trade is possible since no price can be calculated. But Eulox evades this problem by simply putting in the single bid into the formula for price calculation.

This insight is significant because efficiency is critical. If not all possible trades are made, income is forgone, and consequently, Eulox' critical mass of trading volume shifts upwards. Furthermore, Eulox' fee is higher than it would have to be to generate the aimed margin. If creation of trading volume becomes a matter of price, Eulox cannot cut its fee back as much as would be possible with an efficient mechanism. It is questionable if the seeming (marketing) advantages of 'fair and better prices' justify forgoing income in a highly competitive network economy without starting from the pole position.

3.2.4 Conclusion

Table 3.1 shows the essential characteristics of Teleroute, Benelog and Eulox. A similar scheme can be found in Schneider et al. (2002). A considerable number of different electronic transportation marketplaces exist today. Three major categories define how they work: the participants, the traded goods and trade itself. The way that trade is organized – the market mechanism – plays a particularly important role. The market mechanism has implications for smooth transactions and for fees charged. On the other hand, it has to measure up to the market structure and the goods traded.

Teleroute, Benelog and Eulox perfectly demonstrate that Internet-based transportation marketplaces are not interchangeable clones. The three marketplaces also illustrate how the landscape of Internet-based transportation marketplaces perpetually changes. While Teleroute's market model mirrors the traditional ecosystem of carriers, forwarding agents and shippers, Benelog and Eulox try (tried) to utilize actual possibilities and to anticipate future developments. Teleroute's market model has met the past needs but gets increasingly outdated. Eulox had developed a visionary market model – but, this model did not measure up to the structure of transportation markets and the distribution of market power.

	Teleroute	Benelog	Eulox
Participants			
Host	neutral	neutral	seller-sided
Admission	public; restricted	open	open
Structure	many-to-many	many-to-many	many-to-many
#Industries	horizontal	horizontal	horizontal
Traded Goods			
Mode	road haulage	road haulage	road haulage
Terms	short (& long)	(& long)	short
Region	Europe	Europe	Ger/F/A
Requirements	standard non-standard; complete partial	standard & non-standard; complete partial	standard complete
Trade			
Mechanism	blackboard	rev. auction	exchange
Initiative	buyers (& sellers)	buyers	buyers & sellers
Transaction	offline	online	online
Fee	monthly fee; pay per view	transaction-based volume; sellers pay	transaction-based both sides pay
Cooperation	no	no	no

Table 3.1. Teleroute, Benelog, and Eulox.

For the moment, Benelog seems to offer an appropriate compromise between future trends and actual necessities.

Part II

Freight Auctions

4

Conventional Freight Auctions

"Auction design is a matter of 'horses for courses, **not** 'one size fits all'."(Klemperer (2002))

This chapter defines the markets in focus and explores their shortcomings. It also identifies desirable properties of a mechanism that fits these markets better than conventional procedures.

4.1 Markets Considered

From now on, transportation *spot* markets will be considered. These markets are characterized by the fact that mainly single short-term orders are placed by many different shippers. The Internet-based marketplaces in question are those like Benelog, i.e. marketplaces

- that are open,
- where many sellers and many buyers trade short-term transportation capacity by means of buyer-initiated auctions,
- where online transactions are possible.

Table 4.1 describes the considered marketplaces with regard to all aspects of the classification scheme given in the previous chapter. Teleroute, Benelog and Eulox demonstrate that operating freight marketplaces usually try to adapt their mechanisms to the market. But one important aspect has been ignored up until now: transportation markets are markets on which many heterogenous items are traded, and due to so-called *empty lanes* the values of these items are often interdependent. Potential bidders have preferences that are non additive with respect to combinations of items. Conventional placement procedures – posted offers, auctions or even double auctions – do not account for this fact.

	Considered Markets
Participants	
Host	arbitrary
Admission	open
Structure	many-to-many
#Industries	arbitrary
Traded Goods	
Mode	road haulage
Terms	short (preferably)
Region	arbitrary
Requirements	complete
Trade	
Mechanism	auctions
Initiative	buyers
Transaction	online
Fee	arbitrary
Cooperation	yes

Table 4.1. The transportation markets in focus.

4.2 Empty Lanes

Trucks and pick up cars represent the transportation capacities used for road haulage. As a rule, these vehicles have a fixed location where they are maintained and where drivers begin and end their job. Consider a carrier who is located in H. If he has to move some load from a location A to a location B, a determined vehicle must

- first drive from H to A to pick the load up,
- then carry the load from A to B and
- finally return from B to H.

The lanes from H to A and from B to H represent lanes on which the vehicle is moved unloaded. Such lanes are called *empty lanes*. Table 4.1 depicts all lanes necessary for the fulfillment of an order from A to B.

Since empty lanes cause nearly the same costs as lanes on which the vehicle is moved with full loads, empty lanes represent a significant cost driver and as such a major problem for carriers. If, as Figure 4.1 may indicate, the pick-up

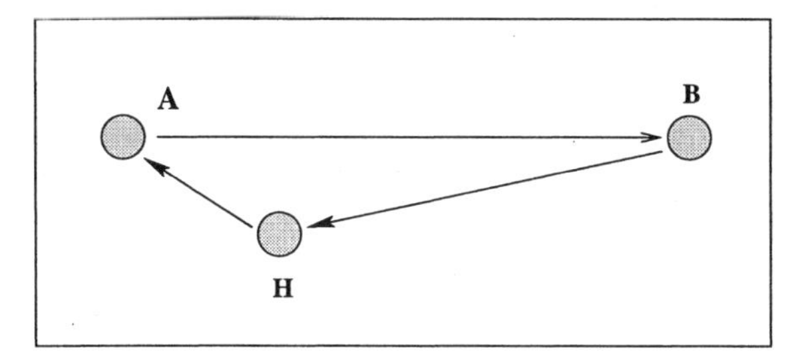

Figure 4.1. Lanes necessary for order fulfillment.

location A is relatively close to H, a carrier can reduce his costs significantly by acquiring an 'appropriate' order from B to A. 'Appropriate' means that both orders can be fulfilled sequentially by one and the same vehicle. Such orders must fit with respect to their pick-up dates, delivery dates and vehicle configuration. Henceforth, two such transportation orders will be called *(mutually) complementary orders*. Carriers will usually try to acquire an order and a complementary order together to maximize capacity utilization.[1] These mutually complementary orders are what the remainder is about.

4.2.1 Standard Auctions & Cost Calculation

For convenience, the following considerations are confined to auctions. Posted offers (as deployed by Teleroute) and double auctions (as deployed by Eulox, for instance) cause the same problem. Suppose that a transportation order o from location A to location B is put up for auction. For ease of exposition only carriers located in A are considered. Moreover, the following assumptions about the transportation market are supposed to hold:

1. An excessive supply of transportation capacity exists.
2. For a given lane, all carriers face the same operating costs. No matter which direction, these costs come to 500 Euro with loaded vehicle and 300 Euro with empty vehicle.
3. Each carrier aims at a total profit of 200 Euro per round trip.

Every carrier must now determine the minimum price for which he would accept o. Table 4.2 depicts different bids of two different carriers X and Y and the subsequent profits or losses

- in case that a complementary o' could be acquired (so that lane BA *is not* empty)
- and in case it could not (so that lane BA *is* empty).

[1] It is obvious that the above consideration also holds for the case where H is close to B. The operating costs directly accountable to the considered transportation order can significantly be reduced by the acquisition of a complementary order from H to A.

Carrier X has no complementary order. Hence o incurs operating costs of 800 Euro. Furthermore, o must earn all the aimed profit of 200 Euro. Thus carrier X is able to bid down to 1000 Euro for o. On the other hand, carrier Y has a complementary order o'. For Y, o incurs accountable operating costs of only 500 Euro. Furthermore, only part of the aimed profit of 200 Euro must be earned with o, say 100 Euro, because the other 100 Euro will be earned with his order o'. Putting this all together, the carrier Y is able to bid down to 600 Euro for o.

Carrier X	Carrier Y	Profit without complementary order	Profit with complementary order
1000	1000	1000 – 800=200	1000 – 500=500
900	900	900 – 800=100	900 – 500=400
800	800	800 – 800=0	800 – 500=300
700	700	700 – 800=–100	700 – 500=200
600	600	600 – 800=–200	600 – 500=100

Table 4.2. Bids in standard freight auctions, profits, and losses.

When bidding for a transportation order, carriers usually do not know whether the acquisition of another complementary order will be successful. So the decision how low to bid is a decision under uncertainty. If a carrier bids like X, his bid is not competitive. If he bids like Y, his bid is competitive but also risky: first, the carrier takes the risk of 200 Euro loss for an eventual profit of only 100 Euro. Second, if he is indeed awarded o for 600 Euro in the actual auction, he cannot bid lower than 600 Euro in later freight auctions in which a complementary order is put up for placement. Consequently, he is not able to underbid competing carriers in later auctions who already acquired a complementary order or who take the acquisition for granted. This means that winning o implies no future advantage.

As a result, the respective 'prognostic abilities' of each carrier are crucial for his economic survival. A sensible minimum bid should be between the bids for the events 'o granted, o' granted' and 'o granted, o' definitely not available'. The most straightforward approach for carriers is to take the safe costs for o and the expected costs for o' for calculation. If, for instance, a complementary order o' can be acquired with a probability of 0.5, the minimum bid in the above example will be 800 Euro. If no such o' can be acquired, a carrier will cover only his costs but he will earn nothing. Correct estimations of probabilities provided, positive profits should be gained in the long run.

> "Now 'in the long run' this is probably true ... but this long run is a misleading guide to current affairs. In the long run we are all dead. Economists set themselves too easy, too useless a task if in

tempestuous seasons they can only tell us that when the storm is long past the ocean is flat again." (Keynes (1923), p. 65)

The expected cost approach fails since losses of earnings represents a substantial problem for carriers with small fleets in particular. Due to longer distances, empty lanes gain even more importance in the USA – so much so that the press stated:

"One of the biggest problems is 'conditional deals': a carrier will agree to move containers in one direction only if it can find someone who will pay it to bring them back again. Unless online auctions accommodate this kind of problem, they will be ignored." (Economist (2000))

Putting the above considerations together, standard auctions do not fit transportation spot markets because they do not account for the problem of empty lanes. This is why combinatorial auctions are considered in the literature as an appropriate mechanism for placing transportation orders.

4.2.2 Combinatorial Auctions – No Solution to Spot Markets

Table 4.3 shows the bids in a combinatorial auction for two complementary orders o and o'. Beside the two carriers X and Y from the previous section a third carrier, Z, takes part in the auction. While Y has already an order from B to A, Carriers X and Z have not acquired any orders yet. Suppose that Z expects to get an order from B to A and that Carrier X does not believe in the acquisition of other orders at all. The rows in Table 4.3 show how X,Y, and Z bid. The last row for instance represents the bid of carrier Z, who demands

Carriers	Order o	Order o′	Total price
Carrier X	x		1000
Carrier X		x	1000
Carrier X	x	x	1200
Carrier Y		x	600
Carrier Z	x		800

Table 4.3. A combinatorial auction for two complementary orders o and o'.

800 Euro for fulfilling order o. The third row contains a bid of carrier X and says that if (and only if) he is awarded both orders his price will be 1200 Euro. Otherwise, his price is 1000 Euro for order o (row 1) and 1000 Euro for order o' (row 2). As described in Chapter 2, the combinatorial auction allocates both orders to carrier X since this allocation incurs the minimal total price. However, there is a necessary condition for the use of combinatorial auctions: there must be *one* shipper offering *many* orders simultaneously. In transportation spot markets, the above orders o and o' typically belong to two different shippers. Consequently, a combinatorial auction cannot be applied

and two separated standard auctions have to be conducted. With the carriers and all bids in the above example (except row 3), o would be allocated to carrier Z and o' to carrier Y. From a global perspective, this is inefficient because total costs would be 1400 Euro.

Summarizing, a packagewize placement of complementary orders reduces uncertainty for carriers. They can submit lower bids without shrinking margins. Shippers in turn benefit from better prices. Hence it is desirable for all market participants. However, up until now, a packagewize placement is not possible in transportation spot markets because in these markets many single orders are placed by many different shippers.

4.3 Packagewize Placements

In fact, there are two major difficulties for packagewize placement:

- Two shippers with complementary orders do have to know about this and they do have to get into contact with each other. In a highly dynamic business like the transportation business with short times for coordination, this alone is a serious obstacle.
- The hammer price for an order package will be a *package* price and the involved shippers have to agree about their respective shares.[2]

With Internet-based transportation markets, the first problem is a rather technical problem. Complementary orders must be aggregated to packages prior to auctioning. What is needed – apart from high trading volume – is an automatic identification and aggregation of complementary orders.[3]

The second difficulty is more subtle. A rule must be established how to divide the resulting package price between the two involved shippers. The following section deals with the requirements to be met by an appropriate division of the package price.

4.3.1 Imbalanced Flows of Goods

> "Bargaining, however, is a costly way to determine prices in societies where time is especially valuable." (Milgrom (1989), p. 19.)

[2] Recall that the price at which an item is knocked down is called the *hammer price*, (cp. Ashenfelter (1989)).

[3] An according software routine has been developed and implemented at DaimlerChrysler's Department for Information and Communication during the project VTE.

Non-automated personal negotiation about the respective shares is time consuming and therefore problematic. Prior to placement, savings are unknown – hence a shipper's readiness for investing the time to search a possible partner and, if found, haggle with him, is certainly limited. Time consumption is a critical issue in transportation business, so much so that posted offers are commonly used for placing transportation orders. Bargaining costs usually outweigh bargaining savings (cp. Bretzke et al. (2001)). Additionally, direct negotiation may also give rise to problems concerning privacy or externalities. Consequently, direct personal negotiation should be bypassed.

The first idea for such a bypass that springs to mind is to equally divide the price. But this approach has a severe shortcoming. It does not account for imbalanced flows of goods, a second peculiarity of transportation markets. The German capital Berlin provides a good example. Typically, more cargo is shipped to Berlin than from Berlin to the outer area. A shipper who has a complete load of potatoes to be carried from Berlin to the surrounding country rarely will be met. Hence such a shipper has a stronger bargaining position than shippers who have potatoes destined for Berlin. A fair mechanism for price division should reflect this fact. Consequently, equal division of the price and with it the savings is generally not fair.

The next idea might be to determine a price division ratio based on the flows of goods. This solution suffers from certain shortcomings as well. First, it is not only one but many different division ratios that are needed. Vehicle requirements, the seize or weight of the cargo and the seasons represent just some of the vital factors for setting such ratios. Furthermore the flows of goods are likely to change by-and-by so that the ratio has to be re-calculated and re-set regularly. Consequently, the effort to establish and update a complete scheme of division ratios is enormous. If not all important factors are accounted for, acceptance problems may occur.

Summarizing, the fixed ratio approach is troublesome. An appropriate rule must lead to a division outcome that is in compliance with the market forces at any given time, so that shippers will accept it.

4.3.2 Monotony, Pareto-Optimality, and Reserve Prices

There are three properties that a reasonable division rule should have: monotonicity (MON), Pareto-optimality (PAR) and compliance with reserve prices, i.e. individual rationality (IR).[4] The latter two properties are obviously necessary. A lack of Pareto-optimality would mean that shippers pay the winning carrier more than they would have to – not terribly realistic. No more realistic would be a division rule that leads to an outcome which charges some shipper

[4] MON, PAR and IR were introduced in Chapter 2.

more than his reserve price for placing his order as is.

Monotonicity is significant as well because the future hammer price H has yet to be discovered. Consider an auction in which an order package is put up for placement that comprises two orders from two shippers S and S'. Figure 4.2 depicts two fictitious bargaining sets that emerge from two successive carrier bids b_1 and b_2. The conflict vector c is determined by the single reservation prices. Carrier bid b_1 gives rise to the negotiable space T_1. Carrier bid $b_2 < b_1$ extends this space to $T_1 \cup T_2$. If the price dividing mechanism was non-monotonous, its graph could be as the bend curve in Figure 4.2. This would lead to the following outcomes: in case $H = b_1$, the division outcome would be given by the intersection of the curve with the first diagonal (i1). But in case $H = b_2$ the outcome is represented by the intersection of the curve with the second diagonal (i2). By means of this division mechanism, shipper S' would be worse off with b_2 than with b_1 – despite the extended negotiable space. If a veto option for the involved shippers was provided, S' would surely not give the green light for carrier bid b_2.

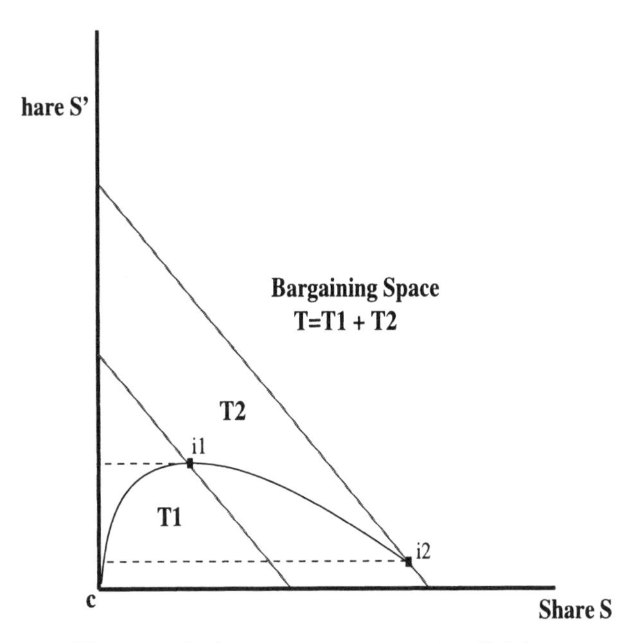

Figure 4.2. A non-monotonous price division.

4.4 Conclusion

The previous sections made clear that a packagewize placement of two complementary spot orders is desirable. Neither standard nor combinatorial freight auctions meet the necessities of transportation spot markets. Consequently, an auction mechanism is needed that

(R1) allows a packagewize placement of complementary spot orders
(R2) provides a division of the resulting package price that is
- (a) driven by the market,
- (b) monotonous,
- (c) efficient,
- (d) individually rational, i.e. in compliance with shippers' reserve prices.

(R3) is safe against collusion and robust against manipulation.

Regarding transportation business, the auction mechanism should also

(R4) be simple and not provide any additional effort for shippers or carriers.

Such a mechanism will be introduced next, and it will be investigated analytically and experimentally afterwards.

5

Dynamic Alliance Auctions

This chapter serves two purposes. First, it briefly explains necessary stages of a mechanism that allows for a packagewize placement of transportation orders in Internet-based marketplaces. Second, it introduces a mechanism that satisfies the requirements **(R1)**-**(R3)** as Chapter 6 will later show. This mechanism is called *Dynamic Alliance auction.* The Dynamic Alliance auction represents a result of the research project *Virtual Truck Enterprise* at the Department for Research and Communication at DaimlerChrysler. To some extent, it has already been explored by Ihde & Schild (2002).

5.1 Stages of an Appropriate Mechanism

A placement of order packages requires that orders that can be packaged do exist and that they will indeed be packaged. In transportation spot markets most orders are single orders that stem from different shippers. Consequently, the desired mechanism must run through the stages of order collection, order aggregation, and placement. The chronological order of these stages is obvious. First, orders must be collected before they can be aggregated to packages that are then put up for placement (cp. Figure 5.1). Additionally, there must be a price division rule that prescribes how much each shipper must pay whose order has been auctioned off in a package.

Collection Stage

In the *collection stage* shippers specify their orders and give all necessary information. All transportation orders specified in this stage are stored in lists.

The collection of orders can be triggered by time or by events. It is triggered by time if it starts and ends at fixed dates. An event-triggered collection

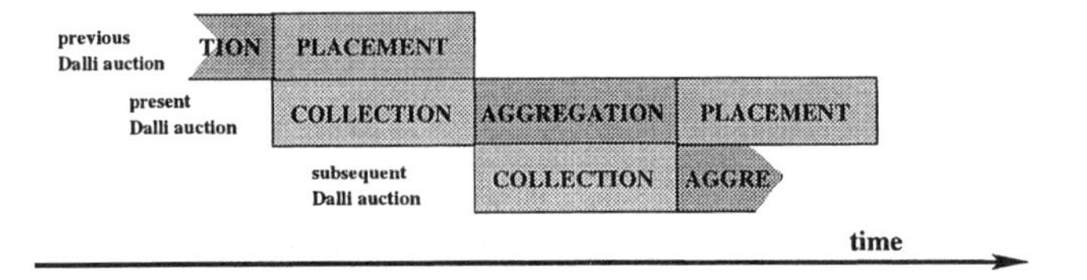

Figure 5.1. Stages of an appropriate mechanism.

ends and starts whenever a certain number of mutually complementary orders is available. If this number is one, for instance, an order is stored until a complementary order arrives. At time this happens, the actual collection stage ends, the two orders are instantly matched, and a new collection stage begins.

Another issue concerns the availability of information about other orders. The collection of orders can either be public or non-public. In a *public collection* all orders specified so far are made publically known. In a *non-public collection* they are kept secret. The distinction between public and non-public collection is important since it gives rise to completely different games. With a public collection stage, shippers enter a sequential game in which the one who waits the longest might be favored. Whether such favoring indeed occurs depends also on both the price division rule and matching rule. On the other hand, a closed collection stage leads to a one-shot game in which shippers act simultaneously and minimizes information available for shippers. Section 6.2 will give more details.

Aggregation Stage

After the collection stage has ended, the *aggregation stage* begins. The hitherto collected complementary orders are tied to order packages (*matched*). If the collection stage delivers s orders from a location A to a location B and s' complementary orders from B to A, there are $s!/(s-s')!$ possibilities for package creation. Consequently, there must be a *matching rule* that describes which two orders will indeed be matched. Such rule could simply say that orders are randomly matched. If no random-based matching is wanted, two obvious criteria for matching are the time when an order is specified by a shipper and the limit set. For instance, FiFMa is a matching rule due to which the order that was first specified will be matched first, so FiFMa represents a *time-based* matching rule.[1] On the other hand, HiFMa represents a matching rule based on limits. It demand that the order with the highest limits be matched first.[2]

[1] FiFMa derives from First in – First Matched.
[2] HiFMa derives from Highest in – First Matched.

Placement Stage

The resulting order packages (and unmatched single orders) are finally auctioned off in the *placement stage*. Either standard auctions or combinatorial auctions can be deployed to place an order package. Here, one faces a trade-off between simplicity and efficiency. While standard auctions are simpler and less time-consuming, combinatorial auctions may be more efficient.

Price Division Rule

A shipper's share of the package price can be determined at some time between matching and end of placement.[3] Section 4.3.1 made clear that the flows of goods must be taken into account for this. However, information on market conditions is not furnished at a certain location, available in explicit form, but spread in the market, as implicit knowledge of shippers. Hence it must be aggregated by means of prices or actions to be taken by shippers. If additional action – like bargaining – is required from shippers, this may be taken in any of the three stages. This depends on the price division rule also.

Interaction of Rules

When designing an auction, there are many parameters to set. When setting them one must take care of the auction as a whole. No stage can be considered separately, and neither can the division rule. It is the interaction of stages and division rule that will determine determines shippers' strategies. For instance, there is no strategic difference between a public and a non-public collection of orders if the matching is random-based. On the other hand, if HiFMa is used, a public collection might favor the shipper who waits the longest (see Section 6.2). Consequently, the new auction format will not perform until all stages work well together.

5.2 How Dynamic Alliance Auctions Work

This section establishes Dynamic Alliance auctions. It starts with an explanation of the basic procedure. Afterwards, it introduces a general notation and gives a compact body of rules for Dynamic Alliance. It concludes with an illustrating example.

5.2.1 The Basic Procedure

Basically, Dynamic Alliance Auctions work as follows.

[3] The absolute price a shipper is charged cannot be calculated until placement.

1. The *collection stage* is non-public and time triggered. Shippers specify their orders and set a so-called *limit*. All transportation orders specified in this stage are stored in order lists. Transportation orders with compatible vehicle requirements and identical pick-up and drop locations are stored in the same list. In practice, the aggregation of the hitherto collected orders should start together with a new collection stage immediately after the collection stage has ended (cp. Figure 5.1).
2. The matching in the *aggregation stage* is based on the HiFMa rule. Every list is ordered by declining limits. The order with the highest limit in a list is matched with the order in the complementary list that has the highest limit. Since for every list exactly one complementary list exists, no ambiguities with respect to lists can occur. The matched orders form packages and their entries are removed from the order lists. This step is repeated until no further matching is possible, which marks the end of the aggregation stage.
3. Finally, all order packages and left-over single orders are auctioned off in the *placement phase.* The rules presented in this section are suited for the deployment of a standard single-good auction. For the usage of a combinatorial auctions, Section 6.4 reports on necessary modifications of Dynamic Alliance Auctions.
4. Package prices are divided between the two involved shippers proportionally to their reserve prices.

5.2.2 General Notation and Terms

This paragraph specifies the most basic variables that will be used throughout the entire thesis. It also defines the situation henceforth in focus.

For ease of exposition, all further considerations are focused on two specific locations, called A and B. All transportation orders from A to B and vice versa are collected in so-called *order lists.* There are two such order lists: $\mathcal{O}$ and $\mathcal{O}'$. All orders from location A to B are stored in $\mathcal{O}$ while all orders from B to A are stored in $\mathcal{O}'$.

There are two different kinds of shippers, those with a transportation order from A to B (i.e. an order in $\mathcal{O}$) and those with an order from B to A (i.e. an order in $\mathcal{O}'$). The set $\mathcal{S}$ is defined as the set of all shippers with an order in $\mathcal{O}$. The set $\mathcal{S}'$ is defined as the set of all shippers with an order in $\mathcal{O}'$.

For ease of exposition, two assumptions are henceforth made.

GA1 Each order in $\mathcal{O}$ is supposed to be mutually complementary with each order in $\mathcal{O}'$.

GA2 No shipper has two mutually complementary orders from A to B and vice versa, i.e. $\mathcal{S} \cap \mathcal{S}' = \emptyset$.

GA1 needs no comment. GA2 is no hard restriction but simplifies matters because a shipper who has two mutually complementary orders could always place them as package on his own. He needs no Dynamic Alliance auction.

5.2.3 The Rules of Dynamic Alliance Auctions

These are the rules for Dynamic Alliance auctions.

1. Collection Stage
   - (Dalli 1a) Beginning and end: Every collection stage begins and ends at an assigned time. Immediately after one collection stage has ended, the next collection stage begins (see Figure 5.1).
   - (Dalli 1b) Participation: To join a Dynamic Alliance auction, each shipper must specify his transportation order and set a *limit*[4]. His order will be stored either in $\mathcal{O}$ or in $\mathcal{O}'$.
   - (Dalli 1c) Non-Public Collection: Order will not be made public in the collection stage.
2. Aggregation Stage
   - (Dalli 2a) Matching: The order with the highest limit in $\mathcal{O}$ is matched with the order that has the highest limit in $\mathcal{O}'$ (HiFMa). The two orders form a package and are removed from the lists. This operation is repeated until one of the lists $\mathcal{O}$ and $\mathcal{O}'$ is empty.
   - (Dalli 2b) Ties: If orders in the same list have identical limits, it is randomly determined which order is matched next, with equal probability for each.
3. Placement Stage
   - (Dalli 3a) Auction Type: Any standard auction for single goods may be used.
   - (Dalli 3b) Reserve prices for single orders: Each order that has not been matched is auctioned off solitarily. Its price limit represents its reserve price in the auction.
   - (Dalli 3c) Reserve prices for packages: Each package is auctioned off with a reserve price equal to the sum of the reserve prices of its single orders.
   - (Dalli 3d) Allocation rule for single orders: The carrier with the lowest bid is awarded the order, provided that his bid does not exceed the reserve price. Otherwise the single order is not awarded to any carrier.
   - (Dalli 3e) Allocation rule for package orders: The carrier with the lowest bid is awarded the order package, provided that his bid does not exceed the package reserve price. Otherwise the order package is not awarded to any carrier.
   - (Dalli 3f) Unallocated order packages: Each order package that has not been allocated to a carrier is dropped.
   - (Dalli 3g) Unallocated single orders: Each single order that has not been allocated is dropped.

[4] The uses of the limits are described in the following rules.

(Dalli 3h) Payment rule for single orders: Each shipper whose single order has been allocated solitarily has to make a payment according to the payment rule of the auction.

(Dalli 3i) Payment rule for package orders: If an order package is allocated to a carrier and H is the hammer price, the two according shippers will jointly have to pay H.

4. Division of the Package Hammer Price

(Dalli 4a) Price division for package orders: The hammer price H for a package order is divided between the corresponding shippers S and S' proportional to their limits. If shipper S set the limit B and shipper S' the limit B', then S must pay $H \cdot B/(B + B')$ while S' is charged $H \cdot B'/(B + B')$.

The interaction of rules will be illustrated by the following example.

5.2.4 An Illustrative Example

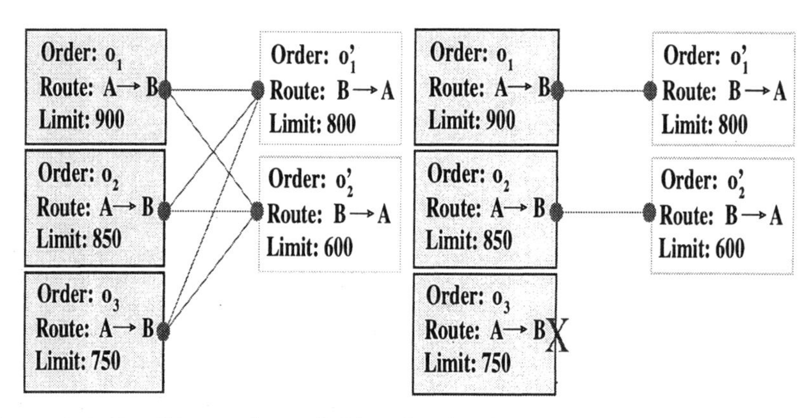

Figure 5.2. Possible matchings (left) and realized matchings, according to HIFMA (right)

Suppose that there are three shippers S_1, S_2, and S_3 in $\mathcal{S}$ with orders o_1, o_2, and o_3 in $\mathcal{O}$ (respectively) and two shippers S_1' and S_2' in $\mathcal{S}'$ with o_1' and o_2' in $\mathcal{O}'$. Suppose that the limits are 900 for o_1, 850 for o_2 and 750 for o_3. The limit for order o_1' is set to 800, and that for o_2' to 600. Figure 5.2 depicts the five orders, each of which symbolized by a square. As the right side of Figure 5.2 shows, 3!/(3-2)! possibilities for packages exist.

Applying the matching rule (Dalli 2a) to the current situation leads to the outcome depicted on the right side in Figure 5.2. Order o_1 is matched with o_1' and order o_2 to order o_2'. On the other hand, order o_3 is not matched. Table 5.1 shows the subsequent reserve prices (second column) and division ratios (third column). The reserve price for the package (o_1, o_1') is calculated according to rule (Dalli 3c), which gives $900 + 800 = 1700$ Euro. The reserve price for the package (o_2, o_2') is $850 + 600 = 1450$. Since order o_3 will be auctioned off solitarily, its reserve price equals its limit of 750. The price shares are calculated according to rule (Dalli 4a). For instance, shipper S1 has to pay a fraction of 900/(900+800) of the hammer price for the package (o_1, o_1'), compare Table 5.1.

Item put up for auction	Reserve price	Division ratio
package order (o_1, o_1')	1700	(900/1700, 800/1700)
package order (o_2, o_2')	1450	(850/1450,600/1450)
single order o_3	750	none

Table 5.1. Matching outcome, the subsequent reserve prices and price division.

After the matching, the respective auctions follow. Suppose that the hammer prices were 1000 for (o_1, o_1'), 1100 for (o_2, o_2') and 600 for o_3. Hence the order packages (o_1, o_1') and (o_2, o_2') are successfully placed because their hammer prices were less than their reserve prices (Dalli 3e). On the other hand, o_3 is not placed because its hammer price exceeds its reserve price (Dalli 3d). Table 5.2 captures the post-auction and therefore final outcome. For each order,

Shipper	Order	Hammer price	Allocated	Charge
S_1	o_1	1000 for (o_1, o_1')	yes	529.41
S_2	o_2	1100 for (o_2, o_2')	yes	644.83
S_3	o_3	600 for o_3	no	0.00
S_1'	o_1'	1000 for (o_1, o_1')	yes	470.59
S_2'	o_2'	1100 for (o_2, o_2')	yes	455.17

Table 5.2. The final outcome of a Dynamic Alliance auction.

it shows whether it could be successfully placed and how much the corresponding shipper has to pay then. Consider for instance, the first row in Table 5.2. It says that order o_1 was successfully placed. The hammer price was 1000 for the order package (o_1, o_1'), so that S_1 is charged $(900/(900+800)) \cdot 1000 = 529.41$.

Part III

Evaluation

6

Stages and Price Division

Chapter 5 introduced Dynamic Alliance Auctions as a mechanism that allows for a packagewize placement of complementary orders in transportation spot markets. Dynamic Alliance auctions are simple and do not require more information or additional actions from shippers than conventional freight auctions. Hence shippers are not provided with extra effort. Hence requirement (R4), formulated in Section 4.4, is already satisfied.

This chapter explores the stages and gives the economic motivation for the proposed design of Dynamic Alliance auctions. The exploration will show that the requirements (R1), (R2), and (R3) are also met. The first section will make a trade-off visible that shippers face when setting their limit. This trade-off is mainly due to the rules for matching (Dalli 2a) and the division of package prices (Dalli 4a). The second section deals with the collection stage and the third with the placement stage. The fourth section examines the aggregation stage while the division of package prices is investigated in the fifth section.

6.1 A Trade-off for Shippers

In Dynamic Alliance auctions, Limits have three functions:

- they serve as reserve price due to (Dalli 3b) and (Dalli 3c),
- they determine which orders are matched (Dalli 2a) and
- they are used for the calculation of price shares (Dalli 4a).

This presents each shipper with the following trade-off. Suppose shipper S has set his limit equal to B. If his order is matched with a complementary order that has the limit B', shipper S's price share of the future package price will be $a(B, B') := B/(B + B')$. Since

$$\frac{\partial a(B, B')}{\partial B'} = \frac{-B}{(B + B')^2} < 0, \tag{6.1}$$

shipper S is interested in getting a partner with a highest possible limit. This means that the higher S's partner's limit is at any given own limit, the smaller his portion of the package price. The probability to get a partner with a high limit increases as S sets a higher limit. This gives S an incentive to set his limit higher.

On the other hand equation (6.2) gives S an incentive to set his limit as low as possible.

$$\frac{\partial a(B,B')}{\partial B} = \frac{B'}{(B+B')^2} > 0. \tag{6.2}$$

Equation (6.2) shows that the lower S's limit is at any given partner's limit, the smaller is his portion of the package price. However, if a shipper sets his limit too low, he might not be matched at all. In this case, he could not benefit from lower prices in package auctions. Moreover, if he sets his limit extraordinary low, there might be no carriers in the subsequent single order auction who underbid the limit. Since limits are the reserve price for single orders, the order is not placed at all. The shipper would have to undertake new efforts for placing it. This is time consuming so that, presumably, shippers will try to avoid such failures.

6.2 Collection Stage

An event-triggered collection stage would automatically induce a time-based matching rule like FiFMa or a random-based matching. Section 6.3 will show that HiFMa is desirable so that a time-based collection has been chosen.

The collection of orders is non public because

- it is proof against collusion,
- it prevents so-called *bid sniping.*

Proofness against Collusion

A look at what Dynamic Alliance auctions do clarifies the eventual aim of collusion among shippers. Dynamic Alliance auctions aggregate orders to packages and split the package price in dependence of the limits set. The trade-off described in Section 6.1 is crucial here. Collusion in Dynamic Alliance Auctions would mean that shippers reach an agreement that aims at minimizing the package price share of shippers (in case they are matched).

Two constellations can be thought of. First, a shipper in $\mathcal{S}$ might try to collude with a shipper in $\mathcal{S}'$. The only thing that could happen is that both shippers decide to build an order package on their own, without Dynamic Alliance auctions. This could hardly be called collusion. Second, collusion might occur between shippers who either all belong to $\mathcal{S}$ or all belong to $\mathcal{S}'$. However, since the collection is non-public, any collusive agreement is not self-enforcing. The argument is the same as in first-price sealed-bid auctions. A re-visitation of the example in Section 5.2.4 and Figure 5.2 shows what can happen. Suppose that shippers S_1' and S_2', who both belong to $\mathcal{S}'$ try to collude. They determined who should get matched first and how to bid. Suppose S_1' and S_2' agreed on the following plan: S_1' bids 580 instead of 800, followed by S_2' who bids 570. Shipper S_2 could set his limit slightly above 581 (to e.g. 582) and would be matched first. His share of the package price would then be 582/1482, which is less than 580/1432. Hence, with a non-public collection stage, Dynamic Alliance auctions are collusion-proof.

Bid Sniping

Bid sniping has been introduced in Chapter 3. In Internet-based auctions that are open and have a predetermined ending time, often the following happens: only a few bids are submitted in the course of an auction, but shortly before it ends, bidders 'jump in' and try to make the last bid. This behavior appears regularly in eBay 2001 auctions, which have fixed deadlines. The initiator of such an auction is worse off than in an auction without sniping (cp. Roth & Ockenfels (2002)). If Dynamic Alliance auctions had a public collection stage, shippers would also have an incentive to join as late as possible. The example from section 5.2.4 will make this clear. Suppose that all shippers except S_1 have already specified their orders and limits. If the collection stage would be public, shipper S_1 knew all other orders and limits. Consequently, he could bypass the trade-off described in section 6.1. For instance, he could set his own limit equal to 851 instead of 900 and would still be matched with S_1'. This way he could reduce his own share from $900/(1700)$ to $851/(601 + 700)$ as Figure 6.1 illustrates.

Of course, this plan succeeds just in case that S_A is the last shipper in $\mathcal{S}$ who places an order with a limit of more than 600. However, the example demonstrates that the shipper who waits longest is favored because he is better informed than other shippers and can, to a certain degree, bypass the trade-off connected with setting a limit. For the proper operation of Dynamic Alliance auctions it is neither desirable that one shipper is favored nor that the outcome gets a gamble because of sniping. Consequently, the use of a public and therefore sequential collection is not advisable.

Summarizing, there are goods arguments why the collection stage should be non-public. Moreover, the experimental results also indicate that information on other orders should not be made available for shippers.

6.3 Aggregation Stage

With s shippers in $\mathcal{S}$ and s' shippers in $\mathcal{S}'$, there are $s!/(s + s')!$ different packages that could be built. Since a shipper's share of a package price does

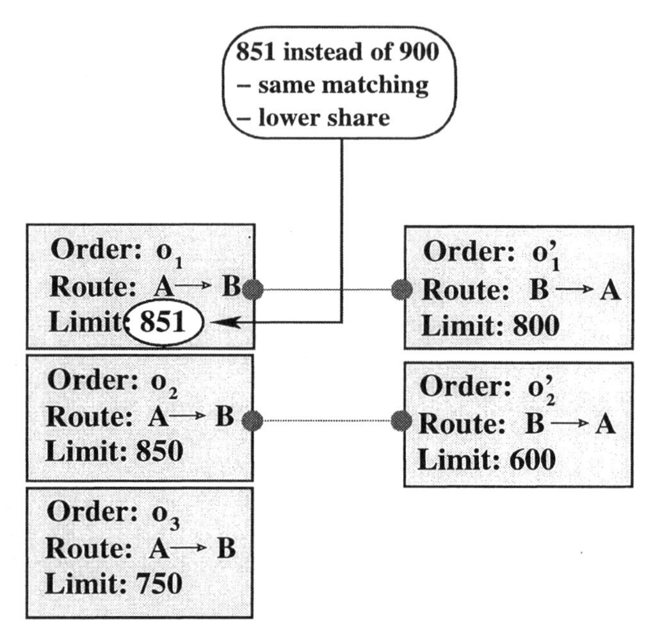

Figure 6.1. The case of complete information.

not only depend on his own limit – but also on the limit of the matched complementary order – the choice of a reasonable matching criterion is crucial. In principle, there are three reasons why HiFMa was introduced as matching rule.

- HiFMa is in compliance with shippers' preferences for matching. In terms of the marriage problem, it leads to a matching outcome that is two-sided stable from an ex-post point of view.
- HiFMa presents an order-preserving rule with respect to matching probabilities.
- HiFMa establishes a market-driven demand aggregation.

These reasons will be explored in the three paragraphs to come.

Compliance with Preferences

Assume that all shippers prefer lower prices for their transportation orders to higher prices. This assumption is very likely to hold in reality. At any rate it is supposed to hold for this paragraph.

Suppose that the collection of orders has just ended and that all orders in $\mathcal{O}'$ (including their limits) have been revealed to the shippers in $\mathcal{S}$. Note that they all have identical preferences on the set of orders available for matching.

This follows directly from the assumption that lower prices are preferred to higher prices and from the fact that a higher the partner's limit is at any given own limit, the lower is one's own share of the hammer price for the package. The most preferred complementary order for matching in $\mathcal{O}'$ is the one with the highest limit. The second preferred order is the one with the second highest limit and so on. Since each order can be matched only once, a conflict of interests exists that must be solved.

On the other hand, all shippers in $\mathcal{S}'$ would have an analog preference ordering if all orders in $\mathcal{O}$ (including their limits) were revealed to them. The most preferred order for matching among all complementary orders in $\mathcal{O}$ is the one with the highest limit. The second preferred is the one with the second highest limit and so on.

Suppose all orders in O and O' are ranked by declining limits as in Figure 6.2. Regarding matching, O can be considered as preference list of shippers in S' and O' as preference list of shippers in S. Consequently, HiFMa matches two orders with the same rank in the lists of preferences (cp. Figure 6.2). Figure 6.2 also illustrates that in case $|\mathcal{O}| \neq |\mathcal{O}'|$, the least preferred orders are those that remain unmatched. The following theorem summarizes this observation.

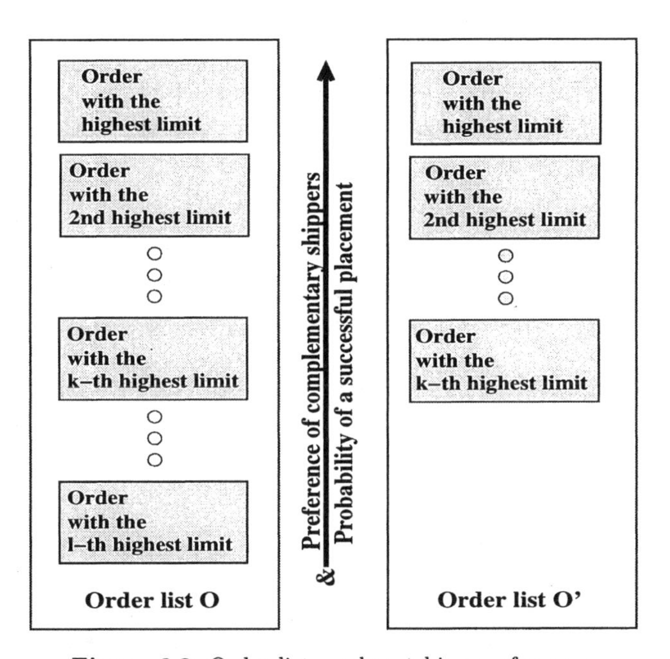

Figure 6.2. Order lists and matching preferences.

Theorem 6.1. *If each shipper prefers a lower price for his order to a higher price, the matching rule (Dalli 2a) in Dynamic Alliance auctions leads to a two-sided stable matching outcome from an ex-post point of view.*

Proof. 'Ex-post point of view' simply abbreviates the supposition that the collection of orders has just ended and that all complementary orders and their limits have been revealed to the shippers. The rest is obvious. □

Order Preservation for Placement Probabilities

Whether an order (package) will be awarded to a carrier depends on the minimum bids of the carriers and the reserve price. This means that orders with high reserve prices are more likely to be successfully auctioned off than orders with low reserve prices. Hence $\mathcal{O}$ and $\mathcal{O}'$ in Figure 6.2 can be also considered as a ranking with respect to the probability of a successful placement. Since the reserve price of a package equals the sum of the reserve prices of its single orders, the reserve price of the package has the same rank as the reserve prices of its single orders. Hence the ranks with respect to the probability of a successful placement remain untouched.

Under the (realistic) assumption that shippers prefer a higher placement probability to a lower probability, a result analogous to Theorem 6.1 can be reached. Concerning shippers' preferences with respect to a successful placement of their order, the matching in Dynamic Alliance auctions leads to a two-sided stable outcome.

Market-Driven Aggregation

Since limits represent the only criterion for matching, the aggregation of complementary orders is market-driven. The similarity to NASDAQ-like stock exchanges is obvious. In NASDAQ exchanges limits present the exclusive matching criterion, too. The highest buy bid is matched with the lowest sell bid. Of course, the similarity stops here. Whereas NASDAQ-exchanges bring supply and demand together, Dynamic Alliance Auctions combine demand (for two different items).

Concluding, neither random-based matching nor time-based rules like FiFMa have one of the above properties. Actually, rules based on time provide shippers with an additional incentive to specify their orders as soon as possible. The outcome of the matching, however, will be similarly arbitrary as with a random-based matching unless the collection of orders is open. The only difference is that shippers have an incentive to place an order as quickly as possible to increase the probability of getting matched.

6.4 Placement Stage

The rules (Dalli 3b) and (Dalli 3c) state that shippers' limits serve also as reserve prices. Due to these rules, Dynamic Alliance auctions impose no extra effort on shippers (requirement (R4)). The rules also links the placement stage with the other stages and the division of package prices. They induce a certain 'bottom line' for the limits since shippers can be assumed to try to avoid a failed auction.

The placement stage is kept quite flexible so that Dynamic Alliance auctions can be used together with any standard auction. Combinatorial auctions could be used as well, but one faces a trade-off between efficiency and simplicity. The rules for placement should be modified and get more complicated than with a standard auction as the following paragraph will show. When developing Dynamic Alliance auctions, a premium was always put on simplicity and practicability. This led to the design proposed here.

Combinatorial Auctions

By means of order collection, automatic aggregation and price division, Dynamic Alliance auction make the submission of combinatorial bids possible. But for that a modification of the pricing rule is advisable. This modification is introduced now.

Table 6.1 depicts the bids submitted by carriers in an exemplary combinatorial auction. The best carrier bid for o is denoted by b_C, the best carrier bid for o' by b'_C and the best carrier bid for the package by $\hat{b}_C$.

Carrier	Order o	Order o'	Total price
Carrier 1	x		b_C
Carrier 1		x	...
Carrier 1	x	x	$\hat{b}_C$
Carrier 2		x	b'_C
Carrier 2	x	x	...
Carrier 3	x	x	...

Table 6.1. A combinatorial auction for two complementary orders o and o', and the best bids that carriers submitted.

In this situation, four conditions can be identified under which a packagewize order placement with the described price division ratio $(b/(b + b'), b'/(b + b'))$ makes sense.

First, the package reservation price must be underbid (equation (6.3)) and the sum of b_C and b'_C must at least equal the best bid for the package (equation

(6.4)):

$$\hat{b}_C \leq B + B' \tag{6.3}$$

$$\hat{b}_C \leq b_C + b'_C. \tag{6.4}$$

Shipper S benefits from a packagewize order placement only if his portion of the subsequent package price is at least as low as the price for a solitarily placement. This can be formalized as

$$\hat{b}_C \cdot B/(B + B') \leq b_C. \tag{6.5}$$

Analogously, shipper S' benefits from a packagewize placement if and only if

$$\hat{b}_C \cdot B'/(B + B') \leq b'_C \tag{6.6}$$

holds. Of course both condition (6.5) and condition (6.6) should be assumed to hold.

An interesting case occurs if equation (6.3) and equation (6.4) hold, but one of the equations (6.5) or (6.6) does not. This means that a packagewize placement of the orders leads to a minimal total price but that one of the two shippers still prefers placing his order alone. Put differently: while a packagewize placement gives rise to the best overall price, it is will not be individually rational in case the price division ratio $B'/(B + B')$ is maintained.

Suppose (6.6) would not hold, i.e. shipper S' would be better off in a solitary placement. How should the orders be placed? In this situation, (6.5) must hold since otherwise (6.3) and (6.4) could not be true, as supposed. Hence a package placement would mean an advantage for shipper S but a disadvantage for shipper S'. Nevertheless, a package placement might still make sense. All it takes is that the monetary benefit of S is big enough to compensate the disadvantage of S' and a revision of the division rule. 'Big enough' simply means that S is neither made worse off by the revised rule than by a solitary placement nor that his limit is exceeded.

Formally, a package bid should be selected as winning bid if

$$b_C \geq \hat{b}_C - b'_C \tag{6.7}$$

and

$$B \geq \hat{b}_C - b'_C. \tag{6.8}$$

Inequality (6.7) holds by assumption (6.4). Thus inequality (6.8) represents a simple criterion to check whether a package placement should be conducted even if (6.6) does not hold. If $B \geq \hat{b}_C - b'_C$, shipper S can be charged $\hat{b}_C - b'_C$ and still benefits from the package placement, while shipper S' faces no longer a loss. Dividing the package price according to $\hat{b}_C \cdot (b_C/(b_C+b'_C), (b'_C/(b_C+b'_C))$ represents another possibility for such a compensation. However, it is consistent with the single reservation prices only if $B > b_C$.

Putting this together, the following modified rules seem advisable if Dynamic Alliance auctions are to be used with combinatorial auctions.

3'.) Placement Stage

(Dalli 3'.a), (Dalli 3'.b), (Dalli 3'.c) These rules are the same as the original rules 3.a), 3.b) and 3.c).

(Dalli 3'.d) Allocation and pricing rule for package orders: The carrier with the lowest bid for the package is awarded the order package for the price of his bid, provided that

1. his bid does not exceed the package reserve price nor the sum of the best bids for the single orders,
2. $B \geq \hat{b}_C - b'_C$ or $b' \geq \hat{b}_C - b_C$ holds.

If the package reserve price is not underbid, the package will not be placed. In case the reservation for one of the single orders was underbid, this order is placed for the corresponding lowest bid.

If the sum of the two best bids for o and o', respectively, is not greater or equal to the best package bid, the two orders will not be placed as a package but solitarily, for the corresponding lowest bids.

(Dalli 3.e') Price Division: If the package is placed, the package will be divided as follows. If (6.5) and (6.6) do hold, price division will be as in (6.5). In case (6.5) or (6.6) do not hold, either $\hat{b}_C \cdot B/(B+B') > b'_C$ or $\hat{b}_C \cdot B'/(B+B') > b'_C$ will hold. If $\hat{b}_C \cdot B/(B+B') > b'_C$, shipper S will be charged b_C while S' pays $\hat{b}_C - b_C$. If $\hat{b}_C \cdot B'/(B+B') > b'_C$, shipper S will be charged $\hat{b}_C - b'_C$, while S' pays b'_C.

With these rules, those bids are awarded the orders that lead to the maximum possible savings for the shippers, including side payments. Carriers profit from a an increased bidding flexibility.

6.5 Division of the Package Price

The price division ratio represents the focal point of this section. Consider two shippers S and S' whose orders have been matched in a Dynamic Alliance Auction so that their orders are auctioned off as a package. What would a price division look like if two shippers tied their orders to a package themselves and bargained their respective shares of the resulting package price conventionally?

A frame of reference to this problem is delivered by the theory of bargaining, in particular by the axiomatic approaches as described in Chapter 2. The (asymmetric) Nash solution (1950), the Kalai-Smorodinsky (Kalai &

Smorodinsky (1975)) solution, and the proportional solution (see e.g. Roth (1979)) represent references for any price division mechanism. This section will show that the price division rule satisfies the requirements (R2)(a) - (R2)(d) from Section 4.4. Furthermore, it will demonstrate that the price division in Dynamic Alliance auctions represents not only a proportional solution but also an asymmetric Nash solution.

6.5.1 Axioms Satisfied

Throughout this section it is assumed that the shippers' utilities for the freight prices are linear and that each of them has a reservation price for placing his order solitarily, say $|B|$ and $|B'|$.

Since B and B' represent maximum expenses, $B, B' < 0$ is assumed to hold. Consequently, $|B + B'|$ is the only individually rational reservation price for the order package. If H denotes the future and at present unknown hammer price for the order package, S and S' face the bargaining game with bargaining set $\{(x, x')|H \geq x + x'\}$ and conflict payoff $\tilde{c} = (B, B')$.

Replacing the hammer price H by $B + B' + s$, where $s > 0$ represents an eventual saving, leads to the bargaining set

$$\tilde{P} := \{(x, x')|B + B' + s \geq x + x'\}. \tag{6.9}$$

The price dividing mechanism used in Dynamic Alliance Auctions has several desirable properties. In particular, it fulfils (R2)(b), (R2)(c), and (R2)(d), as the next proposition will show.

Proposition 6.2. *The price division ratio in Dynamic Alliance Auctions is individually rational (IR), Pareto-optimal (PAR), decomposable (DECO) and monotonous (MON).*

Proof. The price division rule in Dynamic Alliance Auctions belongs to the class of proportional bargaining solutions. Hence it satisfies (IR), (DECO), (MON), and (INV), which were introduced in chapter 2. Proportional solutions lack Pareto-Optimality unless for all elements of the boundary of the bargaining set P, the utility of one player is a declining continuous function of the utility of the other player and vice versa (cp. Holler & Illing (1996)). This is the case in the bargaining problem in Dynamic Alliance auctions because utilities were assumed to be linear. Consequently, PAR is satisfied (cp. Holler & Illing (1996), p. 217). □

Figure 6.3 depicts the bargaining situation in Dynamic Alliance Auctions. It also illustrates monotony and decomposability. The bids $b_1, \dots, b_n$ are decreasing so that the corresponding savings $s_1, \dots, s_n$ (i.e. the bargaining spaces) grow. The arrow contains all price pairs that can occur when the

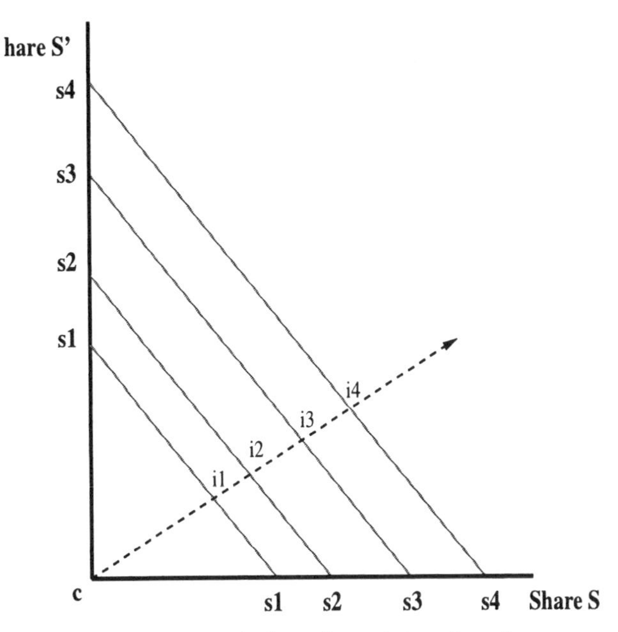

Figure 6.3. Underlying bargaining situation.

hammer price is divided after the auction. With reservation prices B and B', its slope is $(B'/(B+B'))$. The interceptions with the diagonals $(i_1, \ldots, i_4)$ show the division of the actually submitted bids by the Dynamic Alliance auction.

6.5.2 Nash and Kalai-Smorodinsky

The considered bargaining problem is symmetric-equivalent. By means of the order preserving linear utility transformation $\tilde{T} = (T, T')$ with $T : \mathbb{R} \to \mathbb{R}$, $T(x) := (x - B)/s$ and $T' : \mathbb{R} \to \mathbb{R}$, $T'(x) := (x - B')/s$, the bargaining problem (P, c) is transformed to

$$\tilde{T}(P) = \{(x, x') | 1 \geq x + x', x, x' \geq 0\} \tag{6.10}$$

with the conflict payment

$$\tilde{T}(c) = (0, 0).$$

For symmetric-equivalent games, the Nash solution (Nash (1950)) and the Kalai-Smorodinsky solution (Kalai & Smorodinsky (1975)) deliver identical outcomes, namely equal division of the surplus. This holds for the ideal point of Kalai and Smorodinsky as well as the ideal point of Roth, which were both introduced in Chapter 2 (cp. Holler & Illing (1996)). Hence a first result can be fixed.

Proposition 6.3. *Price division in Dynamic Alliance auctions leads to the same outcome as the symmetric Nash solution and the Kalai-Smorodinsky solution if and only if* $B = B'$.

Proof. Dynamic Alliance auctions lead to the symmetric outcome $(B + s/2, B + s/2) = f^N(P, (B, B)) = f^K(P, (B, B))$ if and only if $B = B'$. □

Proposition 6.3 supports the appropriateness of Dynamic Alliance auctions – after all, equal division of the surplus makes sense if and only if both shippers assess the market and thus their bargaining power as balanced. This is exactly what the solution concepts of Nash, Kalai-Smorodinsky and Roth require and where they are congruent with the Dynamic Alliance price division. Generally however, flows of goods are imbalanced so that shippers with orders for the less frequently ordered direction have a stronger bargaining position than the others. This must be accounted for in an appropriate model – equal division represents no plausible option. In fact this is what the asymmetric Nash solution models. Consequently, the asymmetric Nash solution provides a better reference for the Dynamic Alliance price division than the other solution concepts. Theorem 6.4 clarifies the relation between the price division in Dynamic Alliance auctions and the asymmetric Nash solution.

Theorem 6.4. *The price division in Dynamic Alliance Auctions represents an asymmetric Nash bargaining solution* f^w *with weight* $w = B/(B + B')$.

Proof.
Let $w := (B/B + B')$. With respect to IIA and IR the equality $f^w(\tilde{P}, \tilde{c}) = f^w(P, \tilde{c})$ holds, in which

$$P := \{(x, x')|B + B' + s \geq x + x', x \geq B, x' \geq B'\}. \tag{6.11}$$

Since $\tilde{T}(P)$ is the convex hull of the points $(0, 0), (0, 1)$ and $(1, 0)$, the asymmetric Nash solution for this bargaining problem is[1]

$$f^w(\tilde{T}(P), \overline{0}) = (w, 1 - w). \tag{6.12}$$

Due to axiom INV

$$(w, 1 - w) = f^w(\tilde{T}(P), \overline{0}) = \frac{1}{s} \cdot \left(f^w(P, \tilde{c}) - (B, B')\right) \tag{6.13}$$

follows, which is equivalent to

$$(s \cdot w + B, s \cdot (1 - w) + B') = f^w(P, \tilde{c}). \tag{6.14}$$

So the asymmetric Nash solution with weight w imposes expenses of $s \cdot w + B$ on shipper S and $s \cdot (1 - w) + B'$ on shipper S', respectively. Because of $1 - w = B'/(B + B')$ the theorem has been proved. □

[1] Compare to Chapter 2.

6.5.3 An Appropriate Weight

The proposed price division satisfies a series of desirable axioms as Proposition 6.2 showed. Hence it can be considered as a 'fair' bargaining solution. But apart from efficiency, these axioms are fulfilled solely because all proportional and/or asymmetric Nash solutions satisfy them. In principle, every other price division vector $(a, 1-a)$, such that $0 \leq a \leq 1$, would also be individual rational, monotonous, decomposable, independent of irrelevant alternatives and independent of equivalent utility transformations. This is true even for the extreme cases were the chosen a equals 0 or 1. In these cases, one shipper has to pay all of the hammer price alone while the other gets his load moved for free. Such an outcome will be considered appropriate only under special assumptions, e.g. assumptions like in Proposition 7.3.

Axiomatic theory of bargaining offers no assistance at this point because it solely focuses the division mechanism itself – but not the parameters, which are crucial for a concrete case. The main challenge is to find an appropriate weight for dividing the package price. Three arguments suggest that $B/(B+B')$ is the right candidate.

First, the division concept is comfortable and straightforward. Using reserve prices for price division imposes no extra effort on shippers. Calculating their shares in the proposed manner is quite obvious and natural. Each shipper is charged a sum proportional to his share of the package reservation price. Consequently, the approach can easily be communicated to shippers, a necessary precondition for real-life usage.

Second, the price division ratio $(B/(B+B'), B'/(B+B'))$ is the only ratio that satisfies (R 2d) even for the limit case where the hammer price equals $B+B'$. In other words, $(B/(B+B'), B'/(B+B'))$ is the only ratio that is *always* compliant with the reservation prices for the single transportation orders. The reasoning is as follows. Obviously, $(B/(B+B'), B'/(B+B'))$ leads to a individually rational division for both shippers if the hammer price is less than $B+B'$. But now consider the case where the hammer price equals $B+B'$. Assuming individual rationality, shipper S will not accept any share a such that $1 > a > B/(B+B')$ because then S would have to pay $(B+B') \cdot a > (B+B') \cdot B/(B+B') = B$, which is more than his reservation price. Likewise S' will not accept any share exceeding $B'/(B+B')$. On the other hand, neither S nor S' can expect any smaller share since the maximum share of the hammer price is $1 = B/(B+B') + B'/(B+B')$. This proves the statement.

Market-Driven Price Division

Recall that the flow of goods from location A to location B and vice versa should be reflected in the division of the package price. The flows are balanced if there are as many transportation orders from A to B as from B to A.

They are imbalanced in case that one direction has significantly more transportation orders than the other. If the flows of goods are accounted for, the following effect occurs: the more the flows shift from total balance to total imbalance, the more should the division ratio change from (1/2,1/2) to (1,0) or (0,1). A convergence statement would be desirable. Such a statement would consider the price division vectors $(w_{|S|,|S'|}, 1 - w_{|S|,|S'|})$ and say, for instance, that $w_{\infty,|S'|}$ converges to zero as $|S'|$ increases. Analytical evidence that Dynamic Alliance auctions behave this way seems hardly deliverable. However, (1/2,1/2) should be expected for price division for balanced flows of goods, and (1,0) or (0,1) in case flows are imbalanced. Section 7.3 treats this issue. It establishes a simple model for investigation and leads to satisfactory results. In this model, $(1/2, 1/2)$ and $(1, 0)$ are equilibrium division ratios for package prices, as expected. Putting this all together, $(B/(B + B'), B'/(B + B'))$ provides an appropriate ratio for price division.

6.6 Summary

This chapter showed that the Dynamic Alliance auction is a mechanism that fulfills each of the requirements (R1), (R2a), (R2b), (R2c), R(2d), (R3) and (R4). It showed that the stages and price division are closely interrelated and gave the economic motivation for the proposed design.

Issue	Proposed design	Purpose / Property
Collection	time-based non public	prevent random-based matching prevent collusion prevent sniping
Aggregation	HiFMa	two sided stable matching market-driven matching
Placement	standard auctions (can be changed)	simplicity existing platform can be used
Price Division	$B/(B + B')$	market-driven proportional bargaining solution asymmetric Nash bargaining solution

Table 6.2. Explored design issues.

Table 6.2 summarizes the investigated issues. The collection stage is non-public in order to prevent collusion and sniping. The placement stage offers a high degree of flexibility since standard auctions can be used as well as combinatorial auctions (with minor advisable modifications). The aggregation stage uses HiFMa as matching rule. Together with the proposed rule for price division, HiFMa lead to a two-sided stable matching. Furthermore, HiFMa ensures a market-driven demand aggregation. The division of package prices represents a proportional solution and coincides with an asymmetric Nash bargaining solution f^w with weight $w = B/(B + B')$, where B and B' denote the limits of two matched shippers.

7

Efficiency, Payoff, and Bids

This chapter investigates the performance of Dynamic Alliance auctions. First, appropriate performance criteria are introduced in Section 7.1 and a modified private value framework is given in Section 7.2. Sections 7.3, 7.4, and 7.5 explore Dynamic Alliance auctions alongside this framework. Section 7.3 treats two special cases and shows that Nash equilibria with plausible outcomes do exist for them. Afterwards, Section 7.4 develops a formal representation of shippers' expected payoffs, which is used in Section 7.5 in order to demonstrate that the special case equilibria do not hold for arbitrary markets. Intuitive bidding strategies are also offered in Section 7.5.

7.1 Why Investigate Efficiency, Payoff, and Bids

Dynamic Alliance auctions were developed for DaimlerChrysler's Internet-based freight marketplace Fleetboard (2002). The aim was to provide this marketplace with an unprecedent service – an added value to help it reach a high trading volume, which is crucial in a network economy (cp. Shapiro & Varian (1999)). The fee model for DaimlerChrysler's freight marketplace is based on transactions. For each order being placed, the winning carrier is charged a percentage fraction of its hammer price. Consequently, revenue calculates according to

$$\text{revenue} = \text{number of transactions} \cdot \text{transaction fee} \tag{7.1}$$

Taking the transaction fee as given, Dynamic Alliance auctions may lead to a higher number of transactions and thus to a higher revenue. Figure 7.1 shows important factors by which Dynamic Alliance auctions influence the number of transactions (i.e. revenue): simplicity, robustness against manipulation, 'transaction efficiency' and shippers' savings. The following paragraphs explain how.

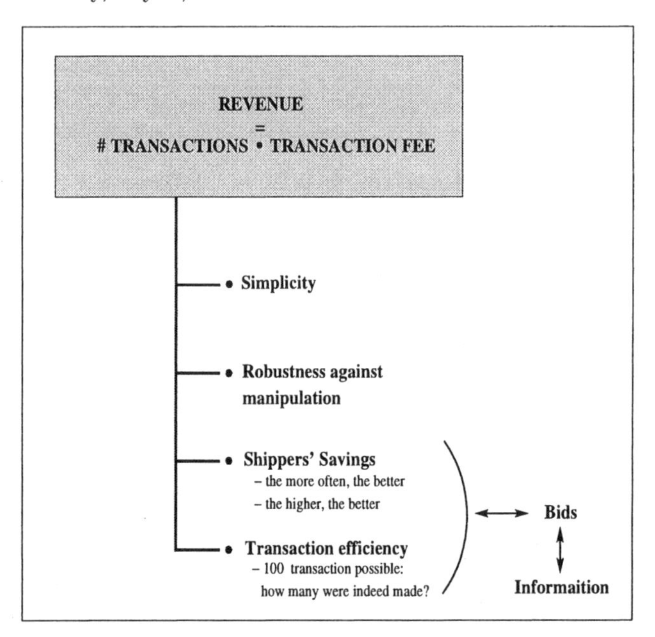

Figure 7.1. How Dynamic Alliance auctions impact the revenue of the marketplace.

Simplicity and Robustness

Simplicity and robustness against manipulation represent obvious and essential prerequisites for real-world usage. It is unlikely that a great number of shippers would use a procedure that is complicated or one where they could be nobbled by unfair practices. Hence the number of transactions could be negatively affected if a complicated or manipulable mechanism was used. Both topics have already been discussed in the previous chapter. Dynamic Alliance auctions are simple and impose no extra effort. Manipulative actions to be considered in Dynamic Alliance auctions would be collusion and, eventually, bid sniping. Section 6.2 showed that Dynamic Alliance auctions are satisfactory in these regards.

Shippers' Savings

Dynamic Alliance auctions were developed to lower carriers' transportation costs through a reduction of empty lanes. This will presumably lead to lower bids and to lower prices for shippers – i.e. to savings in comparison to conventional freight auctions.[1] Shippers' savings by means of Dynamic Alliance auctions will presumably impact the number of transactions. Here, two different aspects are important. The first is how often shippers gain savings. The

[1] Put differently, shippers' payoff will presumably be higher in a Dynamic Alliance auction than in a conventional auction.

more often they save money, the more often they will probably use the marketplace for placing future orders. Additionally, new shippers may be attracted. Consequently, the number of placed orders grows and with it the revenue.[2] The second aspect is how much shippers save per transaction. The higher the savings per transaction are, the more often they will place them using this marketplace, and the more likely it is that new shippers will be attracted. Again, the number of orders grows and with it the revenue of the marketplace (cp. Figure 7.1).

Transaction Efficiency

Figure 7.1 shows that *transaction efficiency* is also important. Transaction efficiency can be defined as the quotient between the number of observable transactions and the number of transactions that would have been possible from an ex-post point of view. In other words, if there are 100 transactions possible, how many transactions will be made indeed? Obviously, transaction efficiency influences the revenue of the marketplace – the higher it is, the more transaction fees can be collected.

However, note that, at least in principle, negative 'payoffs' may occur. In the context of transportation business, *payoff* can be defined as transportation order value minus transportation costs, which is the definition henceforth adopted. In common freight auctions, transportation costs equal the hammer price if the hammer price is less than the reserve price and zero otherwise. The transportation order value deserves a closer look. A shipper is generally someone who manufactures some product and sells it. The price he can charge for this product must cover all of his costs, including the costs to transport the product from his plant to his customer. Hence the natural upper bound for what a transportation order is worth to him should be the difference between the product's price and all costs except for transportation. From now on, this difference will be referred to as his *transportation order value* or simply his *value*. In a Dynamic Alliance auction, a shipper is endangered by a loss only if he sets his limit higher than his value (cp. Section 6.5). His transaction will be called *loss-free* if neither he nor the other involved shipper suffers a loss. With these definitions, a *corrected transaction efficiency* can then be defined as the fraction of observed loss-free transactions and (ex-post) possible loss-free transactions. In other words, if there are 100 orders such each shipper can principally afford to place his order, how many orders will be placed without a loss? Note, however, that the assumption that shippers' values are always higher than the usual market price for their transportation needs might also be reasonable because someone who sells goods but who is permanently unable to cover transportation costs will surely have to go out of business sooner or

[2] The higher number of orders in turn will lead to more matchings and a higher frequency of savings, so that there is a kind of upward spiral.

later. Under this assumption, the two concepts of transaction efficiency and corrected transaction efficiency coincide.

A Remark on Efficiency

In the literature, a standard auction is called efficient if the bidder with the highest value wins the item. In Dynamic Alliance auctions there will generally be more than just one order package. A straightforward extension of the above efficiency concept would be to measure whether those shippers in $\mathcal{S}$ who have the s' highest values are matched when there are $s' < |\mathcal{S}|$ shippers in $\mathcal{S}'$. However, from the viewpoint of the marketplace host, this is not too interesting. Assume, for convenience, that there is one hammer price for single orders and order packages. Without Dynamic Alliance auctions, shippers' accumulated payoff is calculated according to

$$\text{sum of payoffs} \;=\; \text{sum of values} \;-\; \text{number of orders} \cdot \text{ price.} \tag{7.2}$$

With Dynamic Alliance auctions, the following formula is used.

$$\begin{aligned}\text{sum of payoffs} \;=\; & \text{sum of values} \\ & - (\text{ number of orders} - \text{ number of package orders }) \cdot \text{ price}\end{aligned} \tag{7.3}$$

Obviously, the cumulated payoff is greater than with standard auctions if at least one matching occurred. The gain extracted is independent of which shipper was matched but depends solely on the number of transactions. Putting this together, for the host of the marketplace transaction efficiency represents a better criterion than the common efficiency concept.

Bids and Information

Both transaction efficiency and shippers' payoffs depend on the limits that the shippers set (henceforth also called their *bids*). In the simplest case, all bidders bid their value. With such truthful bidders all possible trades would be made so that (corrected) transaction efficiency would be 1 and the cumulated payoff would be maximal.[3] However, for payoff-maximizing bidders, this would be no sufficient argument to bid truthfully. If it seems promising to set the limit differently, there is a good chance that bidders will do so and thereby negatively affect both transaction efficiency and payoffs. In case shippers bid less than their value, e.g. in an effort to reduce their price share of an eventual package, it could happen that not all possible transactions are realized so that transaction efficiency might decrease. This leads to a payoff of zero in situations where a positive payoff could be reached, which in turn reduces the

[3] Here and throughout the following chapters, 'bidders' and 'shippers' will be used as synonyms.

cumulated payoff. If, on the other hand, bidders bid more than their values, e.g. in an effort to maximize their chances to get matched, they might make a negative profit. Again transaction efficiency goes down and with it cumulated payoff. This observation shows that both transaction efficiency and payoff depend on the way that shippers will bid and demonstrates how closely they are interrelated. Nevertheless, these performance criteria are not completely redundant. If two different Dynamic Alliance auctions were conducted and both had a transaction efficiency of, say 90%, the cumulated payoff could be significantly different. For instance, the first Dynamic Alliance auctions could miss transactions where shippers have relatively low values but, for some reasons, the second transactions where values were relatively high. Cumulated payoffs would then be higher in the first auction.

Section 6.2 showed that a shipper's bid will depend on his information about other bids, which was why the collection stage should be non-public. But this might not be the only relevant information. Shippers' bids can also be assumed to depend on information about the number of shippers in $\mathcal{S}$ and $\mathcal{S}'$ in current or past Dynamic Alliance auctions. The effect on transaction efficiency and payoffs is not obvious. Since the marketplace benefits from a high transaction efficiency, the question whether such information should be provided or not is intriguing and will be examined in the last chapter.

7.2 Paul

Consider an exemplary bidder, say *Paul*. What profit can Paul expect? How should he bid? And how efficient will the Dynamic Alliance auction be then? These are the questions that will now be investigated in a (slightly modified) private value model. In fact, the following assumptions are made.

AS1 All shippers in $\mathcal{S}$ and $\mathcal{S}'$, respectively, are identical.
AS2 All shippers in $\mathcal{S}$ and $\mathcal{S}'$ are risk-neutral.
AS3 Shippers' values are independent and identically distributed, according to a uniform distribution on the interval $[0, M]$, $0 < M$, which is commonly known.
AS4 Each shipper knows his own value exactly but not the value of any other bidder.
AS5 The total number of shippers, $|\mathcal{S}| + |\mathcal{S}'|$, is commonly known.

7.3 Polar Cases

This section investigates two 'polar cases' within the modified private value framework and identifies equilibrium bids and subsequent outcomes in Dynamic Alliance auctions. These polar cases can be interpreted as simple situations where the flows of goods between A and B are either balanced or imbalanced. It turns out that there are equilibrium strategies and that either $(1/2, 1/2)$ or $(1, 0)$ are equilibrium division ratios for the package prices. Moreover, transaction efficiency is 1 and the accumulated payoff is maximal.

7.3.1 Perfectly Balanced Flows

The flows of goods between A and B are balanced if there are as many transportation orders from A to B. Consequently, each shipper will be matched with probability 1. In the current setup, balanced flows mean that $|\mathcal{O}|=|\mathcal{O}'|$ holds. To avoid complex notation, each shipper is assumed to have only one transportation order, so that $|\mathcal{S}|=|\mathcal{S}'|$ holds. Define $N := |\mathcal{S}| = |\mathcal{S}'|$.

Four additional assumptions will be made at the beginning since it is not clear which impact shippers' values have on the 'balance' of the situation.

(DV) Shippers' valuations are discrete.
(TB) For each value $x \in [0, M]$ there is exactly one shipper S_x in $\mathcal{S}$ whose value is x.
(TB') For each value $x \in [0, M]$ there is exactly one shipper S'_x in $\mathcal{S}'$ whose value is x.

These three assumptions lead to a symmetric situation. The flows are balanced with respect to both the number of transportation orders and the values of shippers, so that this situation can be considered *perfectly balanced*. Assumptions TB and TB' will be dropped again later. To keep the model as simple and clear as possible, it is further assumed that

(HP) There is one price H for all single and package orders.

Assumption HP is not terribly restrictive. Recall the discussion of empty lanes in Chapter 4. The price for an order from A to B must also cover the costs that the involved carrier needs to get back from B to A. The same holds for orders from B to A. Since empty lanes incur nearly the same costs as full lanes, this assumption captures the spirit of the problem of empty lanes. The vital assumption is that the price for the order package is less or equal to the sum of the prices for the single orders (i.e. subadditive). With the above assumptions, first results can be fixed. To improve the readability of the following sections, some notation is introduced.

Notation 7.1.
Define $h := H/2$.
For all $c \in [o, M]$, *define the strategy* B_c *through*

$$B_c(x) := \begin{cases} x \text{ if } x < c \\ c \text{ if } x \geq c. \end{cases} \tag{7.4}$$

Proposition 7.2. *DV, TB, TB', and HP are supposed to hold and to be common knowledge to shippers. Suppose all shippers know the hammer price* H. *Then* B_h *forms a symmetric Nash equilibrium.*

Proof. See Appendix A. □

The equilibrium in Proposition 7.2 is not the only Nash equilibrium. Consider the family of strategies $\tilde{\mathcal{B}}_h$ that comprises all strategies B such that $B(x) = h$ in case $x \geq h$ and $B(x) < h$ in case $x < h$.

Corollary 7.3. *Under the assumptions of Proposition 7.2, each $B \in \tilde{\mathcal{B}}_h$ forms a symmetric Nash equilibrium.*

Proof. Obvious. □

This multiplicity represents no major problem as the remainder will show. To make a start, all of the above equilibria coincide on values $x \in \{h, \ldots, M\}$. They are symmetric for these values and give all rise to the same division ratio. Consequently, the following corollary can be formulated.

Corollary 7.4. *In any of the equilibria in Proposition 7.2 and Corollary 7.3, the division ratio is* $(1/2, 1/2)$.

The question comes up whether there are any further Nash equilibria. Three observations show that such equilibria cannot be symmetric with respect to values in $\{h, \ldots, M\}$. Consider eventual Nash equilibria (B, B') with the property

(h-SYM) $B(x) = B'(x)$ for all values $x \geq h$.

First, realize that

- there is no Nash equilibrium (B, B') satisfying (h-SYM) such that $B(x) > h$ holds for some value $x \geq h$.

Assume for contradiction there were such strategy B and value x and that all bidders, including S_x, stick to B. This will bring S_x a positive profit of $x - h$. But he can do better by setting his bid equal to $B(x) - 1$. If there is no other shipper in $\mathcal{S}$ who bids $B(x) - 1$, S_x will still be matched with a shipper in $\mathcal{S}'$ who submitted $B(x)$. Thus his payoff will be

$$x - (B(x) - 1)/(2 \cdot B(x) - 1) \cdot H > x - h. \tag{7.5}$$

If there are already some shippers in $\mathcal{S}$ who have also bid $B(x) - 1$, he will also expect a higher profit. Suppose there are n other shippers in $\mathcal{S}$ who also bid $B(x) - 1$. Then there are also n shippers in $\mathcal{S}'$ who submitted the bid $B(x) - 1$. Hence the rule for ties in Dynamic Alliance auctions is applied. S_x will still be matched with the shipper in $\mathcal{S}'$ who has bid $B(x)$ with probability 1/(n+1). He will be matched to someone who bids $B(x) - 1$ with probability $n/(n+1)$. This leads to an expected profit of

$$x - \left(\frac{1}{n+1} \cdot \frac{B(x) - 1}{2 \cdot B(x) - 1} + \frac{n}{n+1}\right) \cdot H,$$

which is more than $x - h$.

Second, realize that

- there can be no Nash equilibrium (B, B') satisfying (h-SYM) such that $B(x) \geq h$ holds for any value $x < h$.

Assume, again for contradiction, there were such strategy B and value x. If all shippers, shipper S_x included, stick to B, S_x will be matched with a shipper in $\mathcal{S}'$ who also submitted $B(x)$. Consequently, the package reserve price is $2 \cdot B(x) > H$. Hence the order package will be placed, and the profit of S_x will be $x - h$, which is negative. Consequently, S_x could do better by bidding less than h, which would grant him a non-negative profit.

Third,

- there can obviously be no Nash equilibrium (B, B') satisfying (h-SYM) with $B(x) < h$ for a value $x \geq h$.

These three observations can be summarized with the next proposition.

Proposition 7.5. *In any Nash equilibrium (B, B') satisfying (h-SYM) shippers will bid h if and only if their value is at least h.*

Proposition 7.5 is satisfactory although the direct consequence is that there are $(h - 1)!$ Nash equilibria satisfying (B, B') satisfying h-SYM. However, the proposition makes also clear that all of these Nash equilibria coincide on values that are at least half the hammer price. It teaches shippers to bid h if their value is as least h. The Nash equilibria are different just for lower valuations, which can be neglected anyway since they cannot increase transaction efficiency. The next corollary is a straightforward consequence of Proposition 7.5.

Corollary 7.6. *Each symmetric equilibrium in Proposition 7.5 will lead to a transaction efficiency of 1 and the maximal cumulated payoff.*

So far, the focus was solely on the existence and 'uniqueness' of Nash equilibria satisfying h-SYM. Now, equilibria will be investigated that do no comply with h-SYM. Rather, they would have to be investigated – it turns out that no such equilibria exist.

Proposition 7.7. *The assumptions DV, TB, TB', and HP are supposed to hold and to be common knowledge to shippers. Suppose all shippers know the hammer price H. Then no Nash equilibrium exists that does not satisfy (h-SYM).*

Proof. See Appendix A. □

The following theorem summarizes Proposition 7.5, Corollary 7.3, and Proposition 7.7.

Theorem 7.8. *Nash equilibria if flows of goods are perfectly balanced DV, TB, TB', and HP are supposed to hold and to be common knowledge to shippers. Suppose all shippers know the hammer price H. Two strategies B and B' form a Nash equilibrium (B, B') if and only if $B, B' \in \tilde{\mathcal{B}}_h$.*

7.3.2 Perfectly Imbalanced Flows

Perfectly imbalanced flows of goods can be treated shortly. Flows are perfectly imbalanced if either IFG or IFG' is valid.

IFG TB holds and $|\mathcal{S}'| = 1$.
IFG' TB' holds and $|\mathcal{S}| = 1$.

Regarding IFG', the shipper that belongs to $\mathcal{S}'$ seems to have an advantage. Intuitively, he should pay less than his colleagues in $\mathcal{S}$. Let $\tilde{\mathcal{B}}_{\mathcal{H}}$ be defined as the family of strategies B such that $B(x) \geq H$ if and only if $x \geq H$. Let B_0 be the strategy of always bidding zero. The next proposition shows that the above intuition is not misleading.

Proposition 7.9. *The assumptions DV and HP are supposed to hold and to be common knowledge to shippers. Suppose all shippers know the hammer price H. In case IFG holds, (B, B') is a Nash equilibrium if and only if $B \in \tilde{\mathcal{B}}$ and $B' = B_0$. In case IFG' is satisfied, (B, B') is a Nash equilibrium if and only if $B = B_0$ and $B' \in \tilde{\mathcal{B}}$.*

Proof. The proposition follows from two statements. First, any (B, B_0) such that $B \in \tilde{\mathcal{B}}$ represents a Nash equilibrium. Second, no other Nash equilibria exist. Both statements are obvious. □

The next statement is an immediate consequence of Proposition 7.9.

Corollary 7.10. *Under the assumptions of Proposition 7.9, the division vector is $(1, 0)$ for IFG and $(0, 1)$ for IFG'. In each case, transaction efficiency is 1 and cumulated payoff is maximal.*

Section 7.3 has delivered satisfactory results, but assuming TB and TB' is quite restrictive. It turns out that Dynamic Alliance auctions are more tricky in general markets, markets where TB and TB' are not granted. The next section provides a formal representation of Paul's expected payoff for such markets.

7.4 Expected Payoff

Assume that Paul's value is $x \in [0, M]$. His payoff does not only depend on x and his bid but also on the bids submitted by the other bidders. For the formal representation, assumption HP is supposed to hold in addition to AS1-AS5. Due to AS 1 (assumption of identical bidders), Paul can assume that

- all bidders in $\mathcal{S}$ play the same strategy $B : [0, \infty) \to \mathbb{R}_+$ and
- all the bidders in $\mathcal{S}'$ play the same strategy $B' : [0, \infty) \to \mathbb{R}_+$.

7.4.1 If $|\mathcal{S}|$ and $|\mathcal{S}'|$ Are Common Knowledge

This paragraph develops a formal representation of Paul's expected payoff in case that $|\mathcal{S}|$ and $|\mathcal{S}'|$ are common knowledge. Define $s' := |\mathcal{S}'|$ and $s := |\mathcal{S}|$. Denote the (realized) values of the bidders in $\mathcal{S}'$ by $y_{(1)}, \ldots, y_{(s')}$ such that $y_{(1)} \geq \ldots \geq y_{(s')}$.

Provided that Paul has submitted the j-th highest bid in $\mathcal{S}$, such that $j \leq s'$, his payoff will be given by

$$u_p(x, y, j) := \mathbf{1_H}(B(x) + B'(y_{(j)})) \cdot \left[x - \frac{B(x)}{B(x) + B'(y_{(j)})} \cdot H\right], \tag{7.6}$$

where the indicator function $\mathbf{1_H}(\mathrm{z})$ is defined as 1 if $z > H$ and 0 otherwise. Hence it is $u_p(x, y, j) = 0$ if the sum of Paul's and his partner's bid is less than H.

On the other hand, if Paul's bid is not among the s' highest bids of the shippers in $\mathcal{S}$, his order will not be matched. His payoff will then be calculated according to

$$u_s(x) := \mathbf{1_H}(B(x)) \cdot [x - H]. \tag{7.7}$$

The probability that Paul submits the j-th highest bid, $j \leq s$, from all shippers in $\mathcal{S}$ equals

$$p_{j,s} := \binom{s}{j-1} \cdot F(B^{-1}(x))^{s-j} \cdot [1 - F(B^{-1}(x))]^{j-1}. \tag{7.8}$$

Hence, for given realizations $y_{(1)} \geq \ldots \geq y_{(s')}$, Paul's expected profit is

$$\pi_{s,s'}(B, B', x, y_{(1)}, \ldots, y_{(N-l)}) := \sum_{j=1}^{m} p_{j,s} \cdot u_p(x, y, j) + \left(1 - \sum_{j=1}^{m} p_j\right) \cdot u_s(y).$$

However, Paul does not know the exact values of the bidders in $\mathcal{S}'$. Consequently, he has to make do with the according random variables for values of the shippers in $\mathcal{S}'$, which will be denoted by $Y_1, \ldots, Y_{s'}$.

Let $Y_{(1)}, Y_{(2)}, \ldots, Y_{(s')}$ be the order statistics of $Y_1, \ldots, Y_{s'}$ such that $Y_{(1)} \geq Y_{(2)} \geq \ldots \geq Y_{(s')}$. Let $f_{(j)}$ denote the density of $Y_{(j)}$ for any $j \leq s'$. The expectation of $\pi_{s,s'}$ with respect to $Y_{(1)}, \ldots, Y_{(s')}$ is given by

$$\tilde{\pi}_{s,s'}(B,B',x) := E_{Y_{(1)},\ldots,Y_{(N-l)}} \pi_{N,l}(B,x,Y_{(1)},\ldots,Y_{(N-s)}) \tag{7.9}$$
$$= \sum_{j=1}^{m} p_{j,s} \cdot \int_0^\infty u_p(x,y) f_{(j)}(y) dy + \Big(1 - \sum_{j=1}^{s} p_{j,s}\Big) \cdot u_s(x).$$

The function $\tilde{\pi}_{s,s'}$ describes Paul's expected payoff in case there are s shippers in $\mathcal{S}$ and s' and Paul knows about this.

7.4.2 If Only the Total Number of Bidders is Common Knowledge

Suppose Paul does not know $|\mathcal{S}|$ and $|\mathcal{S}'|$, but that he knows the total number of bidders, which is denoted by N. For all $s \leq N$, let $q_{N,s}$ denote the probability that $|\mathcal{S}| = s$. Then Paul's expected payoff for a fixed number N of bidders is

$$\pi_N(B,B',x) := \sum_{s=0}^{N} q_{N,s} \cdot \tilde{\pi}_{s,N-s}(B,x). \tag{7.10}$$

7.4.3 If the Number of Bidders is Unknown

Naturally, the most general case is given if AS5 does not hold. In this situation, Paul's overall expectation is given by

$$\pi(B,B',x) = \sum_{N=1}^{\infty} q_N \cdot \pi_N(B,B',x),$$

where $q_N := Prob(|\mathcal{S}| + |\mathcal{S}'| = N)$. However, this case is mentioned just for sake of completeness and will not be considered any further.

7.4.4 Maximizing Payoff

Knowing the formula for his expected payoff, Paul would now like to find out which bidding strategy maximizes his payoff. Because this depends on the others' bidding strategies, he searches for a Nash equilibrium.

If $|\mathcal{S}|$ and $|\mathcal{S}'|$are common knowledge, a Nash equilibrium will generally not be symmetric. A pair of strategies $(B^*_{|\mathcal{S}|}, B'^*_{|\mathcal{S}'|})$ must be found such that

$$\tilde{\pi}_{|\mathcal{S}|,|\mathcal{S}'|}(B^*_{|\mathcal{S}|}, B'^*_{|\mathcal{S}'|}, x) \geq \tilde{\pi}_{s,s'}(B, B'^*_{|\mathcal{S}'|}, x) \tag{7.11}$$

and

$$\tilde{\pi}_{|\mathcal{S}|,|\mathcal{S}'|}(B^*_{|\mathcal{S}|}, B'^*_{|\mathcal{S}'|}, x) \geq \tilde{\pi}_{|\mathcal{S}|,|\mathcal{S}'|}(B^*_{|\mathcal{S}|}, B', x) \tag{7.12}$$

hold for any bidding strategies B, B' and any value x. Obviously, a pair of bidding strategies $(B^*_{|\mathcal{S}|}, B'^*_{|\mathcal{S}'|})$ forms a Nash equilibrium for $\tilde{\pi}_{s,s'}$ if and only if the permutated pair $(B^*_{|\mathcal{S}|}, B'^*_{|\mathcal{S}'|})$ is a Nash equilibrium for $\tilde{\pi}_{|\mathcal{S}'|,|\mathcal{S}|}$. A meaningful equilibrium should generally consist of a pair of different strategies B and B' for the bidders in $\mathcal{S}$ and $\mathcal{S}'$, respectively.[4]

On the other hand, if only the total number N of bidders is common knowledge, the Nash equilibrium searched for can be assumed symmetric. Consequently, a strategy B^*_N must be found such that

$$\pi_N(B^*_N, B^*_N, x) \geq \tilde{\pi}_{s,s'}(B, B^*_N, x) \quad (7.13)$$

and

$$\pi_N(B^*_N, B^*_N, x) \geq \pi_N(B^*_N, B', x) \quad (7.14)$$

hold for any bidding strategies B, B' and any value $x \in [0, M]$.

Equilibrium statements like in Section 7.3 would be satisfactory. Section 7.3 showed that bidders make their bid depend on their own value and the usual market price. All Nash equilibria found had in common to prescribe a bid $h = H/2$ if and only if $x \geq h$. This incurred a transaction efficiency of 1. Unfortunately, the Nash equilibria break down when leaving totally (im)balanced markets.

7.5 Polar Case Equilibria and Intuitive Bidding Strategies

7.5.1 Breakdown of the Polar Case Equilibria

Consider a transportation market with $N = 6$ bidders. Suppose their values are independent and identically distributed according the uniform distribution on $[0, 100]$. Let the hammer price be $H = 50$, so that $h = 25$. Suppose that each of the bidders belongs to $\mathcal{S}$ or to $\mathcal{S}'$ with probability $1/2$, respectively. The expected payoff can easily be derived from equation (7.15) by replacing the relevant terms.[5] It is given by

$$\pi_6(B, B', x) = \sum_{s=0}^{6} \binom{6}{s} \cdot \frac{1}{64} \cdot \tilde{\pi}_{6,s}(B, B', x), \quad (7.15)$$

where

[4] 'Generally' means unless $|\mathcal{S}| = |\mathcal{S}'|$ and $H = H'$.
[5] Note that $q_{N,s} = B(N)(s)$.

$$\tilde{\pi}_{6,1}(B, B', x) = \int_0^{100} u_p(x, y) f_{(1)}(y) dy, \quad (7.16)$$

$$\tilde{\pi}_{6,2}(B, B', x) = \sum_{j=1}^{2} p_{j,2} * \int_0^{100} u_p(x, y) f_{(j)}(y) dy, \quad (7.17)$$

$$\tilde{\pi}_{6,3}(B, B', x) = \sum_{j=1}^{3} p_{j,3} * \int_0^{100} u_p(x, y) f_{(j)}(y) dy, \quad (7.18)$$

$$\tilde{\pi}_{6,4}(B, B', x) = \sum_{j=1}^{2} p_{j,4} * \int_0^{100} u_p(x, y) f_{(j)}(y) dy + \sum_{j=3}^{l} p_{j,2} * u_s(x), \quad (7.19)$$

$$\tilde{\pi}_{6,5}(B, B', x) = p_{1,5} * \int_0^{100} c(x, y) f_{(j)}(y) dy + \sum_{j=2}^{l} p_{j,5} * u_s(x), \quad (7.20)$$

$$\tilde{\pi}_{6,6}(B, B', v) = u_s(x). \quad (7.21)$$

If bidders do not know $|S|$ and $|S'|$ (i.e. how they have been allocated to S and S'), they must choose their bid such that equation (7.15) is optimized. On the other hand, if they are given this information, equations (7.16) - (7.21) are relevant. Of course, the case $|S| = |S'| = 3$ resembles perfectly balanced flows of goods the most.

However, the assumptions of Proposition 7.2 do not hold in the considered market, and it turns out that B_{25} will not form a symmetric Nash equilibrium, no matter whether subjects are informed about $|S|$ and $|S'|$ or not. In order to prove this statement, it a counterexample will be given for the case where subjects know $|S| = |S'| = 3$ as well as for the case that subjects are uninformed (equation 7.15).

Figure 7.2 shows Paul's expected payoff for different strategies if all other bidders use B_{25}. The light lines represent Paul's payoff if he also uses B_{25}. The dark lines show his payoff if he uses B_{50}, instead. The left part in Figure 7.2 represents the payoffs in case bidders know that $|S| = |S'| = 3$, while the right part depicts the payoffs in case bidders are uninformed. In both parts, the sections where the dark line is above the light line show the values where Paul can gain from deviation. These are all values between 43 and 100. Table 7.1 and Table 7.2 show Paul's expected payoffs for values between 43 and 48.

All others use	Paul uses	Value					
		43	44	45	46	47	48
$\mathbf{B}_{25}$	$\mathbf{B}_{25}$	7.60	8.04	8.49	8.94	9.38	9.83
$\mathbf{B}_{25}$	$\mathbf{B}_{50}$	7.79	8.55	9.35	10.19	11.08	12.01

Table 7.1. The case of uninformed bidders: Paul's expected payoff for values between 43 and 48 for B_{25} and B_{50} if all other bidders use B_{25}.

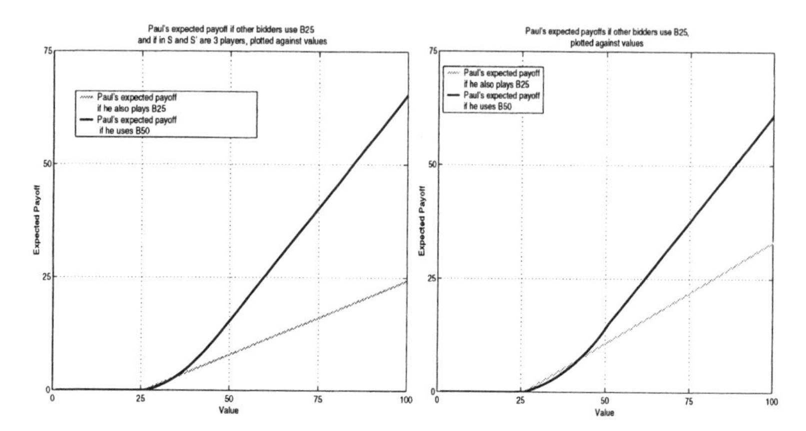

Figure 7.2. A market in which B_{25} forms no symmetric Nash equilibrium. Left: subjects know that $|S| = |S'| = 3$. Right: subjects do not know $|S|, |S'|$.

The expected payoffs have been computed with the numerical software Matlab from equation (7.18) (Table 7.1) and equation (7.15) (Table 7.2). Up until now,

All others	Paul	Value					
use	uses	43	44	45	46	47	48
$\mathbf{B}_{25}$	$\mathbf{B}_{25}$	5.58	5.91	6.23	6.56	6.89	7.22
$\mathbf{B}_{25}$	$\mathbf{B}_{50}$	8.72	9.62	10.54	11.49	12.46	13.44

Table 7.2. The case when bidders know that $|S| = |S'| = 3$: Paul's expected payoff for values between 43 and 48 for B_{25} and B_{50} if all other bidders use B_{25}.

no equilibrium for either case has been detected. So Paul still asks himself how to bid. There are some intuitive ad-hoc strategies he might pursue.

7.5.2 Intuition & Ad-hoc Strategies

A general lesson from Section 7.3 is that bidders make their bid depend on their own value and the hammer price. This insight can be adopted in the general situation.

Consider values lower than the hammer price H. These values will henceforth be called *low values.* A straightforward intuitive strategy for low values lower is to bid the true value (*Intuition* $\mathbf{I}_{low}$). A truthful bid for a low value maximizes the probability of being matched (and getting the order placed) without the risk of a loss.

$\mathbf{I}_{low}$ Bid truthfully for values lower than the hammer price.

Next for values that are at least equal to H (*high values*). As stated, time is money in transportation business. Hence shippers will usually try to avoid failures of placement of an orders with a high value. A bid of at least H for a high value secures the placement of the order, so that no possibility of positive payoff is forgone. Thus it should be quite intuitive to bid this way (*intuition* $\mathbf{I}_{high}$).

$\mathbf{I}_{high}$ Bid at least the hammer price for high values.

Putting the intuitions for 'low' and 'high' values together gives bidders the following advice: bid truthfully for low values and bid at least the hammer price for high values. Of all strategies that obey this advice, there are three strategies prominent: B_{50}, B_{100} and truthtelling (henceforth denoted by B^T).[6] In particular, B_{50} and B_{100} are prominent for the following reason:

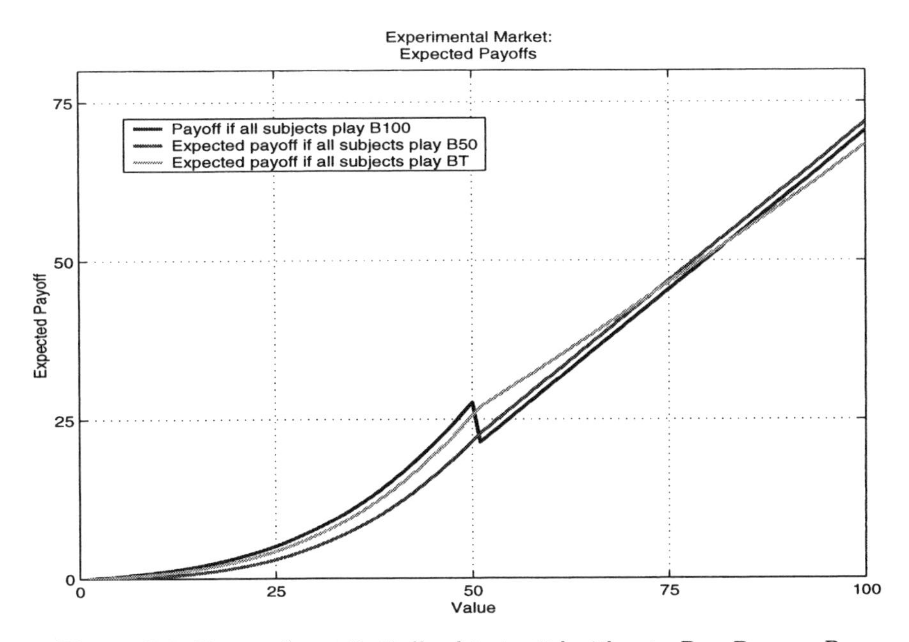

Figure 7.3. Expected payoffs if all subjects stick either to B_{50}, B_{100}, or B_T.

$\mathbf{I}_{50}$ A bid of 50 for high values secures order placement so that no positive payoff is forgone. Furthermore, this bid keeps one's price share minimal in case of a matching.

$\mathbf{I}_{100}$ A bid of 100 for high values secures order placement so that no positive income is forgone. Furthermore, this bid maximizes the probability of being matched.

[6] Recall that for all $c \in [o, M]$, the strategy B_c was defined through $B_c(x) := x$ if $x < c$ and as $B_c(x) := c$ if $x \geq c$ (cp. Notation 7.1).

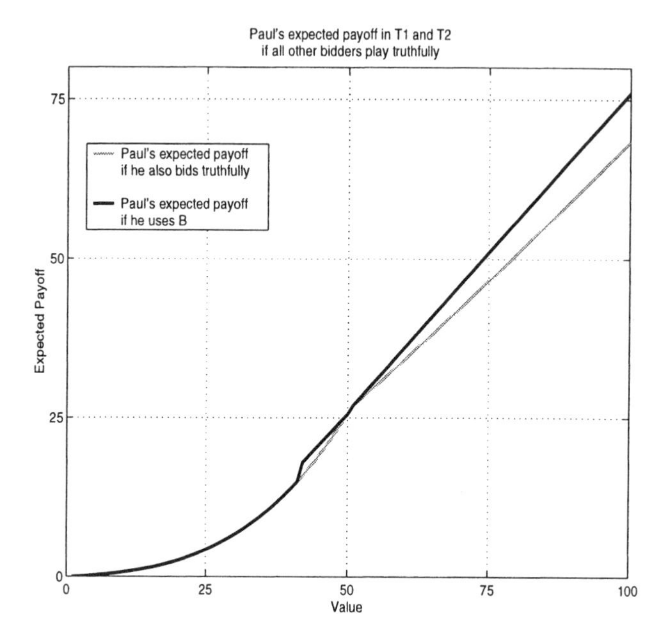

Figure 7.4. Truthtelling forms no symmetric Nash equilibrium.

Figure 7.3 shows the expected payoffs if all subjects would stick either to B_{50}, B_{100} or B_T. None of the intuitive strategies forms a symmetric Nash equilibrium as Figure 7.4, Figure 7.5, and Figure 7.6 illustrate for the case of uninformed bidders. The plots in all these figure are all based on equation (7.15). In each figure, the light line symbolizes Paul's expected payoff if all other player stick to B_{50}, B_{100}, and B^T, respectively, while he uses

$$\widehat{B}(x) := \begin{cases} x & \text{if } x \leq 40 \\ 49 & \text{if } 40 < x \leq 49 \\ 50 & \text{otherwise.} \end{cases} \tag{7.22}$$

The dark line depicts Paul's expected payoff if he also sticks to the respective intuitive strategy. Obviously, for all values between 41 and 49, $\widehat{B}$ provides a better response

- to B_T than B_T itself,
- to B_{50} than B_{50} itself,
- to B_{100} than B_{100} itself

as a look at the figures 7.4, 7.5, 7.6, and Table 7.3 show. Table 7.3 depicts the corresponding expected payoffs, which have been computed with Matlab.

All others use	Paul uses	Value						
		42	43	44	45	46	47	48
$\mathbf{B}_{50}$	$\mathbf{B}_{50}$	13.00	13.92	14.88	15.89	16.94	18.03	19.17
$\mathbf{B}_{50}$	$\widehat{\mathbf{B}}$	13.98	14.93	15.88	16.83	17.77	18.72	19.67
$\mathbf{B}_T$	$\mathbf{B}_T$	15.93	16.97	18.06	19.20	20.39	21.62	22.89
$\mathbf{B}_T$	$\widehat{\mathbf{B}}$	18.01	18.95	19.90	20.85	21.80	22.75	23.69
$\mathbf{B}_{100}$	$\mathbf{B}_{100}$	17.55	18.65	19.80	21.00	22.24	23.53	24.88
$\mathbf{B}_{100}$	$\widehat{\mathbf{B}}$	20.12	21.06	22.01	22.96	23.91	24.86	25.81

Table 7.3. Paul's expected payoff for values between 41 and 49 for various bidding strategies.

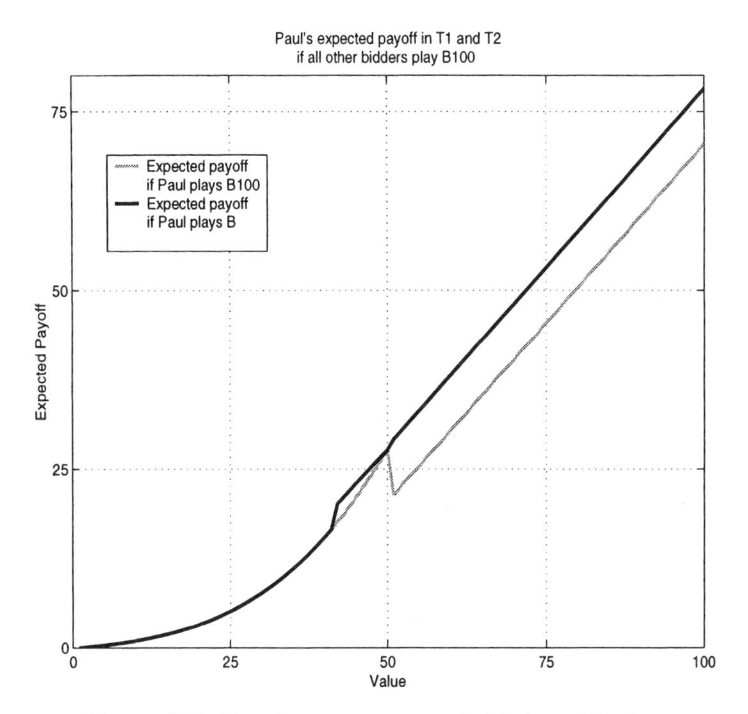

Figure 7.5. B_{100} forms no symmetric Nash equilibrium.

7.6 Summary

Transaction efficiency and shippers savings will usually influence a marketplace's revenue from Dynamic Alliance auctions. Since both depend on the way that shippers bid, bidding strategies have been analyzed under private value assumptions. For the special cases of perfectly balanced and perfectly imbalanced flows of goods, findings are satisfactory. Nash equilibria have been found for such situations, and they give rise to a transaction efficiency of 1. Moreover, the way in which package prices are divided is plausible. The mar-

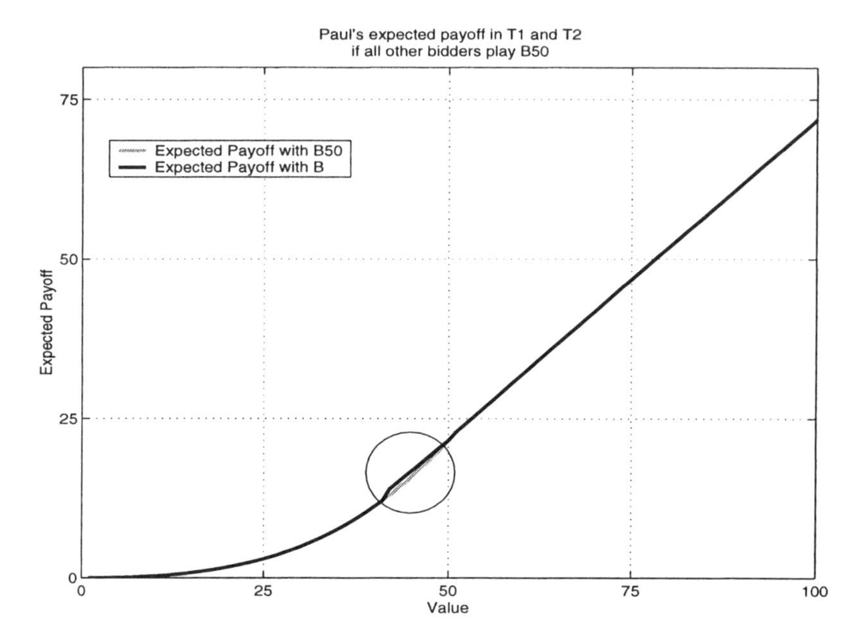

Figure 7.6. B_{50} forms no symmetric Nash equilibrium.

ket with 6 bidders shows that these results are not robust against dropping the special case assumptions. The general function for expected payoff is tricky and Nash equilibria have not been found. Nevertheless, the case of perfect (im)balance provides some intuition for the general case and leads to three prominent ad-hoc strategies: B^T, B_{50}, and B_{100}. To get a deeper insight into bidding behavior and efficiency, the market with six bidders has been experimentally tested. The case in which subjects knew $|\mathcal{S}|$ and $|\mathcal{S}'|$ has been considered as well the case in which only N was commonly known. This is what the next chapter is about.

8

Experiment

The performance of Dynamic Alliance auctions has been experimentally tested, using the market of Section 7.5.1. This market consisted of $N = 6$ bidders whose values were independent and identically distributed according the uniform distribution on $[0, 100]$. Each of the bidders had exactly one transportation order and belonged either to $\mathcal{S}$ or to $\mathcal{S}'$ with probability $1/2$, respectively. The hammer price in this market was $H = 50$ for all orders, no matter whether matched or not. Since transaction efficiency and payoff both depend on the submitted bids and since bidding in turn depends on the information provided, four different informational treatments have been included in the experiment. These four treatments are made up by the two factors 'ex-ante distribution information' and 'ex-post price information' (compare Table 8.1).

'Ex-ante distribution information' means information on the number of bidders in $\mathcal{S}$ and $\mathcal{S}'$ that bidders had been provided with prior to bidding. Because each bidder had exactly one order, this represents the same information as the number of orders from A to B and vice versa. Since it is not obvious how this information impacts the performance of Dynamic Alliance auctions, the question whether a marketplace can enhance this performance by providing such information is intriguing.

On the other hand, 'ex-post price information' means that subjects were informed about how much they were charged and how much payoff they got after bidding. The case that shippers will not be informed how much they have to pay will usually not occur in reality. The same holds for valuations lower than the hammer price. This factor has been included in order to control for eventual spill-overs of bidding due to information feedback. If a subject bids aggressively in Dynamic Alliance auctions, this will have an effect on prices that other subjects must pay. Since these subjects observe these prices, they may draw conclusions ('learn') and change their bidding behavior. This has been evidenced for first-price auctions by Neugebauer & Selten (under revision). Neugebauer & Selten showed that bidding over the risk-neutral Nash equilibrium in first-price auctions, which can be observed frequently, is a result of price information feedback.

		Ex-Post Price Information	
		not given	given
Ex-Ante Distribution Information	not given	T1	T2
	given	T3	T4

Table 8.1. Experimental treatments and provided information.

Table 8.1 shows which treatments were conducted and how they were named. In the first treatment (*T1*), neither price information nor distribution information was provided. Subjects were given ex-post price information in the second treatment (*T2*) but no ex-ante distribution information. The third treatment (*T3*) provided ex-ante distribution information to subjects but not ex-post price information. In the fourth treatment (*T4*), subjects were supplied with both pieces of information. In principle, T2 and T4 can be considered as possible market implementations, and the better one – from the viewpoint of the market participants (shipper, market place host) – should be chosen. T1 and T3 serve as control treatments that cannot be implemented into a market. However, they may allow to assess the feedback influence and to guess how shippers may bid if new on the market, i.e. if not influence by previous market activity.

8.1 Experimental Design

The experiment was conducted in the computer laboratory of the department for BA at the University of Kiel. As a total, 96 subjects were recruited by campus advertisement. All subjects were allowed to take part in one session only. At the beginning of every session, each subject received written instructions for the experiment. Furthermore, the instructions were read aloud to the subjects and the use of the computer software was explained.
As a total, eight sessions were conducted in which the treatments T1, T2, T3, and T4 were tested.

8.1.1 Procedure

In each session two independent runs for the same treatment were conducted simultaneously. Therefore, 12 subjects were allocated to two independent markets consisting of 6 subjects each. Hence two independent sets of observations per session could be collected. For ease of exposition, the remainder considers the experimental procedure for one of the two markets. The procedure for the other market is identical. Each run comprised 100 rounds of Dynamic Alliance auctions. Each order costed 50 units of the experimental currency (ECU).

All rounds proceeded as follows:

1. Each of the 6 subjects in the considered market was randomly allocated to one of two sets S and S' with probability $p = \frac{1}{2}$. This means it was randomly decided whether a subject had a transportation order from A to B or from B to A.
2. Each subject received a private value that was an integer random draw, using a discrete uniform distribution on $[0, 100]$. This value was made known to no other subject. All draws were independent.
3. All subjects submitted their bids, which were restricted to the interval $[0, 100]$. Bidding over one's private value was allowed. In case a subject wanted to bid that way, a warning appeared that this may eventually lead to a loss and a bid confirmation was required.
4. After all bids had been made, the matching took place as described by rule (Dalli 2a) and (Dalli 2b). The subject with the i-th highest bid in $\mathcal{S}$ was matched with the subject with the i-highest bid in $\mathcal{S}'$ (provided that $|\mathcal{S}|, |\mathcal{S}'| \geq i$). The orders of matched subjects were aggregated to an order package.
5. Each package was successfully placed if the sum of the bids of the two matched subjects was at least 50 ECU. The price for each subject was calculated according to rule (Dalli 4a), so that a negative round payoff was possible. Non-aggregated orders were successfully placed if the corresponding subject had bidden at least 50 ECU.
6. Round payoff.

8.1.2 Information

Each subject was informed about his own value, the distribution of valuations and the independence of draws. Each subject was also informed that his round payoff would be the difference between his private value and his price in case his order was successfully placed in that round. All subjects were also told how the matching was conducted and how their prices were calculated. To ensure that all subjects understand that their bids did not only represent maximum prices but impact the matching and price calculation, they were briefly informed about the trade-offs explored in Section 6.1 and taught that all subsequent prices will lie between 0 and 50 ECU.

Depending on the treatment, the following additional information was provided (compare also Table 8.1):

T1 During the experiment, subjects were neither informed about the subject distribution nor about their round prices, round payoffs and total payoffs.

T2 After each round (when all bids had been submitted), each subject was privately informed about his own round prices, round payoffs and total payoffs.
T3 At the beginning of each round (before bids had been made), all subjects were informed about the subject distribution.
T4 At the beginning of each round (before bids had been made), all subjects were informed about the subject distribution. After each round (when all bids had been submitted), each subject was privately informed about his own round prices, round payoffs and total payoffs.

8.1.3 General Remarks

Distributions

Re-consider Paul, the fictitious bidder from the previous chapter. If Paul had been participating in the experiment, in each round he would have faced one of the following situations:

D1 No other subject has been allocated to the same set as Paul (S or S'). Five subjects have been allocated to the other set.
D2 Exactly one other subject has been allocated to Paul's set. Four subjects have been allocated to the other set.
D3 Exactly two other subjects have been allocated to Paul's and three subjects to the other set.
D4 Exactly three other subjects have been allocated to Paul's set, two subjects to the other set.
D5 Exactly four other subjects have been allocated to Paul's set. One subject has been allocated to the other set.
D6 All other subjects have been allocated to Paul's set and none to the other.

D1-D6 will be referred to as the *distributions of subjects.* The corresponding payoff functions are given by equations (7.16) – (7.21). The distributions actually impose three degrees of competitions on Paul and all other subjects as well: no competition, competition and fierce competition. D1 and D6 are totally free of competition since Paul's bid influences neither his matching probability nor to whom he gets matched. D1 guarantees the best possible matching and D6 rules out any matching. D2 and D3 represent competitive situations. Paul's matching probability is still 1 but who Paul is matched to depends on the rank of his bid. D4 and D5 are characterized by fierce competition. Matching is possible but can no longer be taken for granted – it depends on Paul's bid, too.

In T1 and T2, no information on subject distribution was provided. Hence the payoff function is given by equation (7.15), which is the weighted average of (7.16) – (7.21). Furthermore, distributional differences in bidding behavior should be coincidental – Section 8.2.2 confirms this expectation. As a consequence, the analysis of T1 and T2 does not have to be differentiated with respect to subjects' distributions. Things look different with treatments T3 and T4, in which subjects are informed about their distribution prior to bidding. In these treatments, D1, D2, D3, D4, D5, and D6 must be examined separately.

Low Values and High Values

The hammer price of 50 bisects the value interval and gives rise to *low* values (< 50) and *high* values (≥ 50), compare Section 7.5.2. Having a low value represents a somewhat different situation than having a high value because

- with a low value, a bidder cannot afford a placement without a partner, and bidding more than his value might lead to a loss
- with a high value, a bidder can afford a placement without a partner, of and bidding more than his value cannot lead to a loss.

This difference might lead to different bids for high and low values, as the polar case Nash equilibria (Section 7.3) and the intuitive bidding strategies (Section 7.5) already indicate. And indeed, interesting differences can be found in the experiment between data based on low values and data based on high values.[1] This is why a separate exploration of bids (or payoffs) based on low values and those based on high values will generally be pursued.

8.2 Experimental Results

All conclusions presented in this section are based on the outcomes of non-parametric tests.[2] Subjects' payoffs are investigated first, followed by transaction efficiency and bidding behavior.

8.2.1 Average Payoffs and Revenue Efficiency

Table 8.2 reports average round payoffs and *normalized* average payoffs.

The normalization was reached by dividing the average payoffs by the average payoff in T1. The payoff in T1 and T2 is slightly different. Separating low values from high values increased the difference (cp. Table 8.3).[3] The

[1] Cp. e.g. the paragraphs on the average payoffs or on the bid-to-value ratios and competition in T3 and T4.

[2] Siegel (1997) gives a comprehensive introduction and explanation of non-parametric tests.

[3] Recall that 'low values' were defined as all values less than the hammer price, while 'high values' represent all values that are at least equal to the hammer price.

	Average Payoff	Normalized Average Payoff
T1	23.02	1.00
T2	23.27	1.01
T3	21.93	0.95
T4	22.20	0.96

Table 8.2. Average payoff in ECU per round and subject.

	Average Payoff		Normalized Average Payoff	
	high values	low values	high values	low values
T1	39.44	5.42	1.00	1.00
T2	41.40	4.88	1.05	0.9
T3	39.15	4.45	0.98	0.82
T4	39.43	4.48	1.00	0.83

Table 8.3. Average payoff in ECU per round and subject for high and low values.

average payoffs in Table 8.2 and Table 8.3 suggest that subjects are worse off in the treatments where they know the distribution of subjects. For bids based on high values, the differences between T1, T4 and T3 are merely noticeable, but T2 leads to 4% higher average payoffs. For bids based on low values, the average payoff is 7% less in T3 and T4. A non-parametric Mann-Whitney-Test of the hypotheses

($\mathbf{H}_{02}$) Average payoffs are equal in T1 and T2.
($\mathbf{H}_{03}$) Average payoffs are equal in T1 and T3.
($\mathbf{H}_{04}$) Average payoffs are equal in T2 and T4.
($\mathbf{H}_{05}$) Average payoffs are equal in T3 and T3.

confirms significant differences between average round payoffs in T2 and T4 for the high-value section and between T1 and T3 for the low-value section. Table 8.4 reports the corresponding significance levels. The differences in av-

	$\mathbf{H}_{02}$	$\mathbf{H}_{03}$	$\mathbf{H}_{04}$	$\mathbf{H}_{05}$
high values	$p = .553$	$p = .299$	$p = .015*$	$p = .624$
low values	$p = .250$	$p = .049*$	$p = .55$	$p = .192$

Table 8.4. Significance levels of cross-treatments differences in average payoffs according to a Mann-Whitney-Test ($n_{T1} = 12$,$n_{T2} = 6$,$n_{T3} = 12$, $n_{T4} = 6$. * rejected at $\alpha = .05$).

erage payoff might have two different explanations. The first might be that a different number of relatively high bids for low values occurred. The second might be that transaction efficiency is lower in because bids lower than the value were submitted. Both kinds of bids would reduce the average payoff.

	LOSSES			
	T1	T2	T3	T4
cumulation	-102	-597	-70	-585
occurrence	10	76	9	59
average per occurrence	-10.2	-7.86	-7.7	-9.9
average per subject & round	-.085	-.166	-.0583	-.163

Table 8.5. Losses in different treatments.

Bidding more than a low value could result in a negative payoff, i.e. a loss. A more frequent occurrence of losses reduces the cumulated payoff. However, this can be ruled out quickly for T3 and T4. If losses in T3 and T4 were at least partially responsible for the reduced average payoffs, these losses would have to be higher than the losses in T1 and T2. Table 8.8 shows that the opposite is true. Losses sum up to 597 ECU in T2, compared to 585 ECU in T4, and to 102 ECU in T1, compared to 70 in T3 (compare the first row in Table 8.8). Table 8.8 also shows that losses occur more frequently in T2 than in T4 and more frequently in T1 than in T3.

On the other hand, bids less than the underlying private values could lead to higher percentage of periods in which a loss-free transaction was possible but not realized, i.e. to a reduction of transaction efficiency.

	Market						Overall
	M1	M2	M3	M4	M5	M6	Average
T1	100%	80%					89%
T2	104%	98%	102%	97%	100%	88%	98%
T3	95%	85%					90%
T4	95%	78%	81%	87%	89%	86%	86%

Table 8.6. Transaction efficiency in T1-markets, T2-markets, T3-markets, and T4-markets (including transactions with losses).

Transaction efficiency has been defined as the fraction of observed transactions and (ex-post) possible transactions (compare Section 7.1). Formally, transaction efficiency is calculated according to

$$\frac{t_{part} + t_{nopart}}{T_{part} + T_{nopart}}. \tag{8.1}$$

In equation (8.1),

- t_{part} represents the number of transactions that were made by matched subjects,
- t_{nopart} represents the number of transactions made by unmatched subjects,

- T_{part} represents the number of loss-free transactions that were possible for matched subjects,
- T_{nopart} represents the number of loss-free transactions that were possible for unmatched subjects.

Table 8.6 shows how 'transaction efficient' the different experimental markets were. The highest overall transaction efficiency can be found in treatment T2, followed by T1, T3, and T4. Consequently, different transaction efficiencies seem to be the cause for the different average payoffs. However, t_{part} and t_{nopart} include those transactions where losses occurred, so that a more frequent occurrence of losses leads to a higher transaction efficiency so that Table 8.6 is biased in favor of T2 and T4. Consequently, the corrected transaction efficiency, which includes only loss-free transactions, seems to be more appropriate (compare Section 7.1). Formally, corrected transaction efficiency is calculated by

$$\frac{t_{part,noloss} + t_{nopart,noloss}}{T_{part} + T_{nopart}}, \tag{8.2}$$

where

- $t_{part,noloss}$ represents the number of loss-free transactions that were made by matched subjects,
- $t_{nopart,noloss}$ represents the number of loss-free transactions made by unmatched subjects,

while T_{part} and T_{nopart} are defined as before.

	Market						Overall
	M1	**M2**	**M3**	**M4**	**M5**	**M6**	**Average**
T1	98%	80%					89%
T2	97%	94%	99%	95%	95%	86%	94%
T3	94%	83%					89%
T4	92%	76%	78%	86%	86%	84%	84%

Table 8.7. Corrected transaction efficiency in T1-markets, T2-markets, T3-markets, and T4-markets (excluding transactions with losses).

Table 8.7 shows the findings for the corrected transaction efficiency of the different experimental markets, and it confirms the order for (the uncorrected) transaction efficiency: the highest transaction efficiency can be found in Treatment T2, followed by T1, T3, and T4. A one-tailed randomization test shows that the difference between transaction efficiency in T2 and transaction efficiency in T4 are highly significant. The null hypothesis

($\mathbf{H}_{06}$) Average transaction efficiency in T2 equals average efficiency in T4 had to be rejected at $\alpha = 0.01$ (p=0.00216). Conclusions on the other treatments are not possible.

Table 8.8 shows the different frequencies with which (strictly) positive payoffs occur in the different treatments. T2-markets produced 2,483 strictly

	PROFITS > 0			
	T1	**T2**	**T3**	**T4**
cumulation	27,723	85,175	23,496	80,516
occurrence	790	2,483	9/641	2,166
average per occurrence	35.09	34.3	10.85	37,12
average per subject & round	23.1	23.66	22	22.37

Table 8.8. Losses and strictly positive payoffs in different treatments.

positive payoffs compared to 2,166 strictly positive payoffs in T4-markets. The same holds for T1 (790 occurrences) in comparison to T3 (641 occurrences).

Summarizing, the treatments in which subjects knew the distributions led to slightly lower losses than the treatments in which subjects did not know. But a considerable reduction of (the corrected) transaction efficiency overcompensated the savings so much so that the average payoffs were significantly lower in T3 and T4.

8.2.2 Bid-to-Value Ratios

For all treatments, quotients between bids and values (*bid-to-value ratios*) have been calculated and statistically explored. In fact, two different kinds of bid-to-value ratios were used – the bid-to-value ratio of a subject and the bid-to-value ratio of a market.

- The bid-to-value ratio b_j of a subject j is represented by the average fraction of bid and respective private value, taken over all rounds: $b_j := \sum_{i=1, v_{ji} \neq 0}^{100} \frac{b_{ji}}{v_{ji}} / 100$. In this formula, b_{ji} denotes the bid and v_{ji} the value of a subject j in round i.

- The bid-to-value ratio ${}_{\mathcal{M}}b$ for a market $\mathcal{M}$ is the mean bid-to-value ratio, taken over all subjects that belong to $\mathcal{M}$, i.e. ${}_{\mathcal{M}}b := \sum_{j:j \in \mathcal{M}} b_j / 6$.

Bid-to-value ratios are useful. A bid-to-value ratio of 1 indicates that, on average, bidders bid their value. On the other hand, a bid-to-value ratio of less than 1 indicates that, on average, bidders bid less than their values while a bid-to-value ratio above 1 that bidders bid more than their value (on average). In order to ensure independency of observations and to prevent price-based

	b^{T1}	b^{T2}	b^{T3}	b^{T4}
D1-D6	0.87	1.06	0.79	0.79
D1	0.82	1.03	0.58	0.63
D2	0.86	1.05	0.71	0.84
D3	0.86	1.15	0.69	0.83
D4	0.87	0.99	0.84	0.85
D5	0.89	1.06	1.03	0.93
D6	0.85	1.08	0.65	0.71

Table 8.9. Average bid-to-value ratios.

learning, bid-to-value ratios for markets are used whenever T2 and T4 are analyzed. For T1 and T3, bid-to-value ratios for subjects are used. The respective bid-to-value ratios are denoted by b^{T1}, b^{T2}, b^{T3} and b^{T4}.

	b^{T1}		b^{T2}		b^{T3}		b^{T4}	
values	**low**	**high**	**low**	**high**	**low**	**high**	**low**	**high**
D1-D6	0.85	0.88	1.15	0.97	0.75	0.83	0.84	0.76
D1	0.73	0.89	1.13	0.96	0.69	0.62	0.64	0.62
D2	0.88	0.84	1.09	0.98	0.77	0.66	0.95	0.73
D3	0.85	0.88	1.14	0.97	0.68	0.693	0.92	0.74
D4	0.82	0.91	1.34	0.97	0.85	0.833	0.88	0.81
D5	0.91	0.88	1.02	0.96	1.08	0.827	1.02	0.83
D6	0.87	0.83	1.15	0.99	0.54	0.671	0.62	0.82

Table 8.10. Average bid-to-value ratios for low and high values.

Table 8.9 reports the average bid-to-value ratios, and Table 8.10 reports the average bid-to-value ratios for high and low values. Interestingly, T2-subjects bid on average about 6% more than their value (cp. Table 8.9). A look at at Table 8.10 shows that a noticeable difference between the bid-to-value ratios in T2 for low and high values exist. The bid-to-value ratio for high values in T2, b^{T2}, is 0.97. So for high values, T2-subjects report, on average, almost exactly their true values. Furthermore, the average bid-to-value ratio will be located between 0.96 and 0.98 with 95% probability (cp. Table 8.11). This suggests that T2-subjects meet the Dynamic Alliance auctions trade-offs by bidding almost exactly their value (if they have a high value). The relative square deviation of bids from the value is 0.017, which also supports the hypothesis

	T1		T2	
	low values	high values	low values	high values
D1-D6	[.98;1.13]	[.91;1.07]	[.98;1.02]	[.74; 1.01]

Table 8.11. 95%-Confidence intervals for mean bid-to-value ratios in T1 and T2. T1: individual bid-to-value ratios, T2: market-based bid-to-value ratio.

of honest bidding. However, Section 8.2.3 will show that honest bidding plays a certain role in T2 but that mainly, differences average out.

On the other hand, T2-subjects bid approximately 21% *above* their true value if they have a low value. A similar bidding behavior cannot be found in T1, T3 and T4. Here, subjects usually bid clearly less than their value, except for D5. But although subjects indeed bid more than their value in D5, this behavior is more moderate than in T2: approximately 8% in T3 and 2% in T4 (cp. Table 8.10). It is a remarkable result that the average bid-to-value ratio is far below 1.0 in D4 and D5 for the high values, particularly because pursuing the strategy B_{100} would have maximized the chance for a matching. However, such bidding strategy also maximizes the own share (given the other bids), and it seems that subjects do not like that idea. On the other hand, for low values, subjects bid more than their true value in D5, even though this may end up in a real loss and not just in a reduction of profit. This behavior seems somewhat contradictory.

	T3		**T4**	
	low values	**high values**	**low values**	**high values**
D1	[0.32;1.06]	[0.24;1.00]	[0.49;0.90]	[0.44;0.75]
D2	[0.62;0.91]	[0.51;0.80]	[0.61;1.33]	[0.61;0.81]
D3	[0.51;0.85]	[0.56;0.83]	[0.79;1.05]	[0.66;0.83]
D4	[0.74;0.95]	[0.76;0.91]	[0.81;0.96]	[0.73;0.89]
D5	[0.53;1.63]	[0.76;0.89]	[0.66;1.40]	[0.76;0.93]
D6	[0.26;0.82]	[0.48;0.86]	[0.49;0.82]	[0.75;0.92]

Table 8.12. 95%-Confidence intervals for mean bid-to-value ratios in T3 and T4. T3: individual bid-to-value ratios, T4: market-based bid-to-value ratio.

When comparing average bid-to-value ratios in T3 and T4 across distributions (Table 8.10), bid-to-value ratios seem to increase with the number of competitors. The next paragraph investigates this issue.

Bid-to-Value Ratio and Competition in T3 and T4

For each treatment a one-sided Page test was conducted to test the following null hypothesis for low and high values, respectively.

$(\mathbf{H}_{07})$ $\bar{b}_{D1} = \bar{b}_{D2} = \bar{b}_{D3} = \bar{b}_{D4} = \bar{b}_{D5}$.

For T1 and T2, H_{07} must be accepted for both high values and low values. As the involved subjects have no information related to subject distribution, eventual differences between the bid-to-value ratios with respect to distribution are coincidental. This was expected. On the other hand, H_{07} must be rejected for high and low values in T3 and T4 in favor of the alternative hypothesis

(H_{17}) $\bar{b}_{D1} \leq \bar{b}_{D2} \leq \bar{b}_{D3} \leq \bar{b}_{D4} \leq \bar{b}_{D5}$.

with at least one strict inequality in H_{17}. Almost the same holds for T4: H_{07} must be rejected for high values and is 'almost' rejected for low values ($p = 0.0505$). The test results are summarized in Table 8.13. Section 8.1.3

	T1	T2	T3	T4
low values	$p = .57$	$p = .1$	$p = .038^*$	$p = .0505$
high values	$p = .78$	$p = .5$	$p = .004^{**}$	$p < .00003^{**}$

Table 8.13. Results of a one-sided Page test for H_{07}.* rejected at $\alpha = .05$. **H_{07} rejected at $\alpha = .01$. $n_{T2} = n_{T4} = 6$. $n_{T1} = 10$ for low, $n_{T1} = 11$ for high values. $n_{T3} = 9$ for low, $n_{T3} = 10$ for high values.

dealt shortly with the distribution of subjects, and information on this distribution was claimed to represent a measure of how fierce competition for orders was. This information was supposed to influence bids. The rejection of H_{07} has delivered empirical evidence for this suggestion.

To find out which distributions lead to different bid-to-value ratios the following null hypothesis were tested.

(H_{08}) $\bar{b}_{D1} = \bar{b}_{D2}$
(H_{09}) $\bar{b}_{D2} = \bar{b}_{D3}$
(H_{010}) $\bar{b}_{D3} = \bar{b}_{D4}$
(H_{011}) $\bar{b}_{D4} = \bar{b}_{D5}$

The tests used were a one-sided randomization test for related samples for treatment T4 and a Wilcoxon-Signed-Rank Test for treatment T3. Table 8.14 summarizes the findings and reports the p-values.

	T3		T4	
	low values	**high values**	**low values**	**high values**
$\mathbf{H_{08}}$	$p = .013^*$	$p = .508$	$p = .016^*$	$p = .016^*$
$\mathbf{H_{09}}$	$p = .71$	$p = .239$	$p > .05$	$p = .016^*$
$\mathbf{H_{010}}$	$p = .019^*$	$p = .019^*$	$p = .016^*$	$p = .016^*$
$\mathbf{H_{011}}$	$p = .48$	$p = .937$	$p > .05$	$p = .016^*$

Table 8.14. Tests for an increase in average bid-to-value ratios as competition increases. *rejected at $\alpha = .05$. T4: one-tailed randomization test ($n_{T4} = 6$) , T3: two-tailed Wilcoxon-Sign-Rank ($n_{T3} = 12$).

For low values in treatment T4, the only significant difference for occurs between D1 and D2. This suggests that subjects only differentiate between a non-competitive and a competitive situation. If competition occurs, its fierceness seems to be irrelevant for bidding behavior. On the other hand, subjects seem to make one further distinction in the high-value section: between a competitive situation, in which matching is guaranteed (D2 and D3), and a highly competitive situation, in which getting a partner cannot be taken for granted (D4 and D5). Next for treatment T3. For low values, subjects make a difference between non-competitive and competitive situations (rejection of H_{09}) as well as between competitive and highly competitive situations (rejection of H_{011}). For high values, the only difference can be found between competitive and highly competitive situations.

Summarizing, subjects respond to all degrees of competition, as expected. The differences in response between a competitive and a highly competitive distribution depend, however, on both the treatment and the value section.

Comparisons across Treatments

	$\mathbf{H}_{012}$	$\mathbf{H}_{013}$	$\mathbf{H}_{014}$	$\mathbf{H}_{015}$
D1-D6	p= .074			
D1		p=.147	p=.745	p=.016*
D2		p=.033*	p=.574	p=.016*
D3		p=.083	p=.349	p=.016*
D4		p=.419	p=1.00	p=.016*
D5		p=.248	p=.925	p=.016*
D6		p=.065	p=.111	p=.030*

Table 8.15. High values: test results for H_{012}, H_{013}, H_{014} (two-tailed Mann-Whitney-Test) and H_{015} (one-tailed randomization test), tested for high values.

Naturally treatments in which subjects know their distribution prior to bidding should be expected to be different from treatments in which subjects do not know their distribution. Also, treatments that differ by ex-post information on payoff have led to different bid-to-value ratios in some cases (cp. Table 8.10, T1 vs T2 and T3 vs T4, low-value section). Consequently, differences between bid-to-value ratios of the four treatments were investigated. Therefore, the following null hypotheses were tested with a Mann-Whitney-Test:

($\mathbf{H}_{012}$) Bid-to-value ratios in T1 equal bid-to-value ratios in T2.
($\mathbf{H}_{013}$) Bid-to-value ratios in T1 equal bid-to-value ratios in T3.
($\mathbf{H}_{014}$) Bid-to-value ratios in T3 equal bid-to-value ratios in T4.
($\mathbf{H}_{015}$) Bid-to-value ratios in T2 equal bid-to-value ratios in T4.

Null hypothesis H_{015} must be rejected for high values in favor of the alternative hypothesis that bid-to-value ratios in T2 differ from bid-to-value ratios in

T4. T1 and T2 differ also significantly, for the low-value section though. This means that H_{012} must be rejected for low-values in favor to the alternative hypothesis that bid-to-value ratios in T1 differ from bid-to-value ratios in T2 for low values. For high values, H_{012} must be accepted, but the corresponding p-value is small (cp. Table 8.15). No significant difference between T3 and T4 could be detected so that H_{014} must be accepted for all distributions and low as well as high values. Also, H_{013} must be accepted. The exact p-values are shown in Table 8.16 and Table 8.15. Putting this together, it seems that

	$\mathbf{H_{012}}$	$\mathbf{H_{013}}$	$\mathbf{H_{014}}$	$\mathbf{H_{015}}$
D1-D6	p= .038*			
D1		p=.272	p=.886	p=.016*
D2		p=.386	p=.399	p>.05
D3		p=.065	p=.092	p>.05
D4		p=.729	p=.851	p=.016*
D5		p=.954	p=.925	p>.05
D6		p=.065	p=.920	p=.016*

Table 8.16. Low values: test results for H_{012},H_{013},H_{014} (two-tailed Mann-Whitney-Test) and H_{015} (one-tailed randomization test), tested for low values.

- ex-post price information significantly impacts subjects' bids if no other information is available[4]
- if information on distribution and ex-post price information are both available, the information on distribution is of greater impact on subjects bids.

The latter conclusion is supported by the detected differences between T2 and T4: H_{015} can completely be rejected for high values and for D1, D3 and D6 in the low-value section. Interestingly, H_{015} could neither be rejected for low values nor for high values (except for D2). But judging by the average payoffs and efficiency results, there must be differences, but they seem to average out.[5]

8.2.3 Bids in T1 and T2: Intuition Revisited

Each of the strategies B_{50}, B_{100} and B_T represents intuitive strategy but no symmetric Nash equilibrium in the experimental market considered (cp. Section 7.5.2). Nevertheless, there were subjects in T2 that indeed pursued B_{50}, B_{100} and B_T persistently. Figure 8.1 shows the bids of three T2-subjects,

[4] Price information 'helps' subjects: remember that the average round payoff was higher in treatments with ex-post price information than in the treatments without it (cp. Table 8.3).

[5] H_{015} was tested with modes, means and medians instead of bid-to-value ratios. But again, except D2, no differences could be found.

plotted against values. In fact, these three subjects were the only subjects that followed the ad-hoc strategies this impressively.[6]

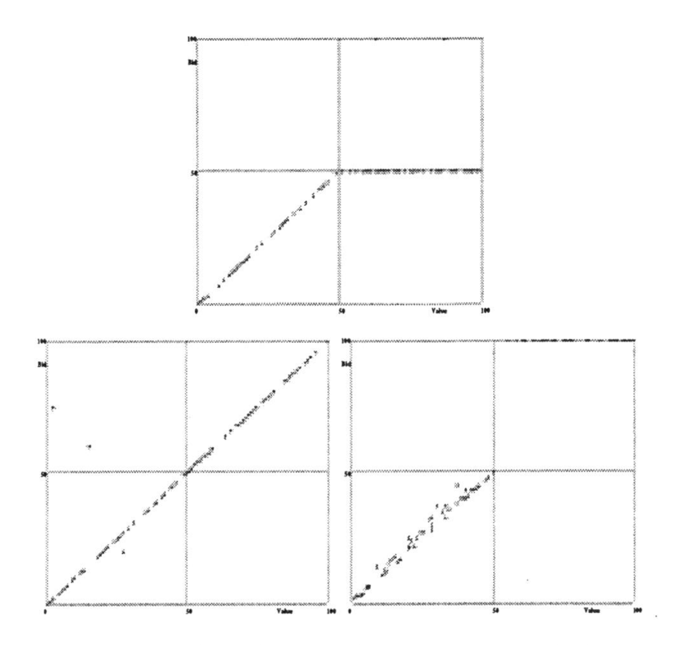

Figure 8.1. Bids of three subjects: B_{50},B_T, and B_{100}.

However, if two relaxations are permitted, many subjects in T1 and T2 can be identified to follow the intuition behind the focal strategies (I_{low}, I_{50}, I_{100} and I_T, cp. page 102). The first relaxation concerns the consistency of bids. Only a predetermined minimum percentage of bids should have to be congruent with a strategy. Threshold levels considered here were 50%, 75% and 90+%. Second, deviations should be allowed. For instance, when considering B_{50}, bids slightly above 50 should also be counted (if based on a high value). This seems reasonable because such bids follow the same intuition as B_{50}: to secure an item if affordable and to keep the own share relatively small in case of a matching. B_{50} is just the most extreme representative for this intuition. Similar arguments hold for B_{100} and B_T, so that deviations should be tolerated for them as well.

Treatment 1

For low values, I_{low} seems to be a strong line of orientation for subjects in T1 (cp. Table 8.17 and Figure 8.2). One third of all subjects did not deviate

[6] One subject in T1 played with a comparably persistent. It deployed B_{35}, which is not really what one could call 'focal'.

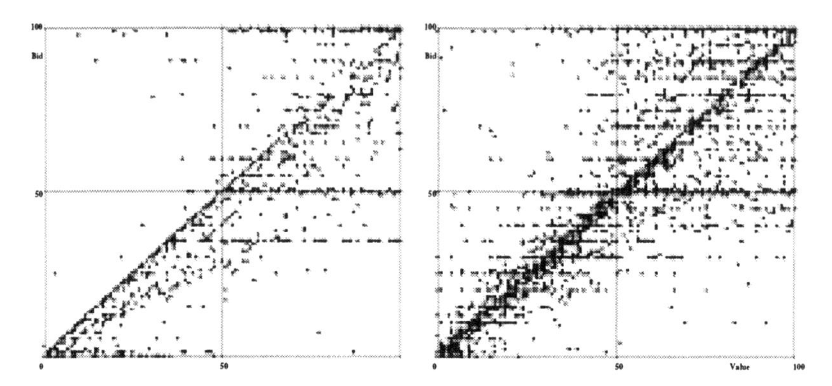

Figure 8.2. Left side: bids in treatment T1, against value. Right side: bids in treatment T2, against value.

from the value by more than 1 ECU (downwards) in more than 94% of their bids. 50% of the subjects bid accordingly in more than 80% of their bids and 58% of the subjects do so for at least 60% of their bids. If deviation occurred, it was almost always downwards, which may be guided by the desire to make a profit worth mentioning and the reflection that the matching probability is not significantly reduced. The high-value section can be treated very shortly. Only one subject pursued B_{50} on a 50% level.

percentage of bids	**>50%**	**>75%**	**>94%**
T1	7(58%)	6(50%)	4(33%)
T2	23(64%)	14(39%)	7(19%)

Table 8.17. Number of subjects whose bids deviated only slightly from the value in T1 (by at most 1 ECU) and in T2 (by at most 2 ECU), in dependence of a required minimum percentage. Total number of subjects was N=12 in T1 and N=36 in T2.

Treatment T2

First for low values. Almost a fifth of all subjects made at least 96% of their bids such that they did not deviate by more than 2 ECU from the true value. More than half of the subjects at least 65% of their bids accordingly,(cp. Table 8.17).

Next for high values. It turned out that many subjects submitted bids between 50 and 55 ECU (I_{50}), between 96 and 100 ECU (I_{100}) or bids that deviated by at most 2 ECU from their value (I_T). As a total, 23 subjects[7] submitted more than 50% of their high-value bids according to I_{50}, I_{100} or I_T, and 16 subjects[8] more than three third of their bids. Table 8.18 gives the details. These findings and Figure 8.2 are also useful to assess the significance of the average bid-to-value ratio for T2, which was 0.99 (cp. Table 8.10). Although honest bidding plays a certain role – after all, a fifth of all bids based on values between 56 and 95 belong to were close to the value – it clearly shows that the bid-to-value ratio is also a result of averaging out.

minimum percentage	**>50%**	**>75%**	**>90%**
$[v-2, v+2]$	8 (22%)	7(19%)	1(3%)
$[50, 55]$	7 (19%)	4(11%)	1(3%)
$[96, 100]$	4 (11%)	3(8%)	1(3%)
$\sum$	23 (64%)	16(44%)	3(8%)

Table 8.18. Number of subjects whose bids deviated only slightly from the value in T1 (by at most 1 ECU) and in T2 (by at most 2 ECU), in dependence of a required minimum percentage. Total number of subjects was N=12 in T1 and N=36 in T2.

8.2.4 Summary

The highest transaction efficiency and average payoff could be found in T2. The differences in transaction efficiency between T2 and T4 were highly significant. Average payoffs could be found significantly lower in treatments where subjects knew their distribution (differences depend on the value section considered). Differences in bidding behavior also existed. Bidding behavior was investigated using bid-to-value ratios, which directly indicate whether subjects bid, on average, their value or less or more than their value. T2 is prominent again. First, T2 is the only treatment where subjects bid remarkably above their value in the low value section. Second, the mean bid-to-value ratio for high values is 1, which could be taken as indication for truthtelling. However, this is just an effect of averaging out. Comparing bid-to-value ratios of different treatments, significant differences between T2 and T4 were detected for high values: T4-subjects submitted significantly lower bids than T2-subjects. In both treatments T3 and T4, different distributions of subjects (D1-D6) lead to significantly different bid-to-value ratios. This evidences that subjects

[7] For 18 subjects, more than 50% of the high-value bids could be assigned to one of the three categories. For 5 subjects, more than 50% of the high-value bids lie in two of the three categories.

[8] For 14 subjects, more than 75% of the high-value bids could be assigned to one of the three categories. For 2 subjects, more than 75% of the high-value bids lie in two of the three categories.

respond to different degrees of competition for complementary orders. Intuitive bidding (according to B_{50}, B_T, and B_{100}) was also considered and found to play a small but noticeable role: at least 8% of all T2-subjects submitted more than 90% of their bids accordingly.

9

Putting Insights to Practice

A new auction-based mechanism has been introduced and evaluated. It allows for a packagewize placement of complementary transportation orders in Internet-based spot markets, it is simple and it imposes no extra effort on shippers. This mechanism is called *Dynamic Alliance auction.* Dynamic Alliance auctions were developed for DaimlerChrysler's Internet-based freight marketplace (Fleetboard (2002)). The aim was to provide this marketplace with an unprecedent service.

Insights Gained I

With respect to the mechanism itself, four results can be gained without a significant theoretical framework. First, Dynamic Alliance auctions are safe against manipulative actions like collusion or bid sniping. Second, they lead to a division of the resulting package price that is driven by the market and belongs to the family of asymmetric Nash solutions for bargaining problems if utilities are linear in money. The rule for price division in Dynamic Alliance auctions satisfies also the axioms of monotonicity and decomposability. Third, orders are aggregated to packages (or matched) in a way that is compliant with shippers' preferences (as long as shippers prefer to pay less for an order than more). In terms of the marriage problem, the matching outcome is two-sided stable from ex-post. Fourth, the mechanism is appropriate for any standard auction format but can also be used for combinatorial auctions as well. A few small modifications will further enhance efficiency.

These results are valuable since they suggest that Dynamic Alliance auctions are suited for real-world application. How well Dynamic Alliance auctions will indeed perform will undoubtedly impact the success of the marketplace. Due to the transaction-based fee model of DaimlerChrysler's marketplace – for each order being placed through the marketplace, the winning carrier is charged a percentage fraction of its hammer price – immediate implications for the marketplace revenue exist. Reasonable performance criteria for Dynamic Alliance auctions are given by shippers' payoffs and 'transaction efficiency'. Transaction efficiency is defined as the quotient between the number of observable transactions and the number of theoretically possible transactions. To study the performance of Dynamic Alliance auctions, further assumptions had to be made.

Model

The assumptions represent a modified private value model, which allowed an analysis and experimental exploration of Dynamic Alliance auctions. Two specific locations A and B were considered and two different sets of shippers: those with a transportation order from A to B and those with an order from B to A. For ease of exposition every order from A to B was assumed to be mutually complementary with every order from B to A. All shippers in were assumed to be risk-neutral and identical (apart from the direction of their transportation order). Their values were assumed independent and identically distributed, according to a uniform distribution that was supposed to be commonly known. Every shipper was supposed to know his own value exactly but not the value of any other shipper. All shippers were also assumed to know the usual market price for orders, which was supposed to be the same for all orders and order packages. The total number of shippers who participate in a Dynamic Alliance auction was also commonly known.

Evaluation

The performance of Dynamic Alliance auctions within this model was investigated both analytically and experimentally. Since transaction efficiency and payoff both depend on shippers' bids and since shippers' bids in turn depend on the information available to them, four different informational treatments were studied in an experiment with 96 subjects. These four treatments are made up by the two factors 'ex-post price information' and 'ex-ante distribution information'. 'Ex-ante distribution information' means information on the number of orders from A to B and orders from B to A. In the first treatment (T1), neither price information nor distribution information was provided. In the second treatment (T2), subjects were given ex-post price information but no ex-ante distribution information. The third treatment (T3) gave distribution information to subjects but not ex-post price information. In the fourth treatment (T4), subjects were supplied with both pieces of information. The intriguing task was to find out how will Dynamic Alliance auctions perform in the different settings. The results for T2 and T4 are more significant for practice since ex-post information about prices will, of course, always be provided in reality.

Insights Gained II

The analytical investigation led to the following insights. First, plausible Nash equilibria exist for some special cases. In all of these equilibria, Dynamic Alliance auctions lead to a transaction efficiency of 1 and the maximum of the possible payoffs for shippers. Also, these equilibria outcomes seem to reflect the relation of orders from A to B and from B to A (*the flows of goods*). In equilibrium, the price for a package order is divided equally between the two involved shippers if these flows are perfectly balanced. It turns out that this Nash equilibrium is not unique - but the strategies of the other equilibria coincide on certain values and lead to the same outcome (concerning allocation and payoffs). In case of perfectly imbalanced flows, equilibrium strategies can also be found. Here, one shipper will pay all while the other gets the transportation for free. The Nash equilibria for the special cases break down in arbitrary markets, but they give an inspiration for intuitive ad-hoc strategies.

The experiment brought the following results. First, the performance of Dynamic Alliance auctions was best if subjects were given ex-post price information but no ex-ante distribution information (T2). Comparing T2 to the treatment with price and distribution information (T4), the difference came to 5% for high values and to 17% for low values. The main reason for this can be seen in a significantly higher transaction efficiency, which was about 94% in T2-markets but only 82% in T4-markets. On the other hand, markets with ex-post price information performed slightly better than markets without this information. Although no significant differences could be detected, it seems that price information helps subjects bid.

Second, it seems that ex-post price information significantly impacts subjects' bids if no other information is available. If, on the other hand, information on both prices and distribution is provided, distribution information seems of greater impact on subjects' bids. This conclusion is based on the observation that bids are different in T1 and T2, and different in T2 and T4, but that there is no significant difference between bids in T3 and T4.

Third, the bids in T4 and T3 can be interpreted as strategic bids that account for the competition for complementary orders. The average quotient of value and bid increased as the number of complementary orders decreased and (since there the market size was constant) simultaneously the number of orders for the same direction increased. The strongest response to that competition could be observed in treatment T4.

Fourth, some T2-subjects pursued one of the intuitive ad-hoc strategies throughout the entire experimental run. However, this was only a small minority. Nevertheless, when allowing deviations for the intuitive strategies, many bids could be found that were located nearby.

Transfer of Insights & Future Research

A market mechanism must fit the market and it must be accepted by market participants. For a successful Internet-based marketplace, acceptance of the deployed market mechanism is vital indeed. Judging by the simplicity, robustness of Dynamic Alliance auctions, this new auction format may gain the necessary support in the market. The market-driven proportional division of the package price should also be easily communicated and be acceptable for shippers. Additional support is provided by the analysis of Dynamic Alliance auctions, which showed that outcomes are plausible for the extreme cases of perfectly balanced and perfectly imbalanced flows of goods.

The experimental insights suggest that a restrictive information policy on current Dynamic Alliance auctions should be adopted. It seems preferable not to provide information how many orders from A to B and vice versa actually are available. It may also be better not to provide a 'history' on this information, i. e. data on past Dynamic Alliance auctions that states how many orders for each direction participated. Such information may lead to strategic bids as in T3 and T4 and result in lower transaction efficiency, lower shippers' savings, and consequently, in a reduced marketplace revenue.

Analysis and experiment, however, are based on several assumptions, which provided a starting point for investigations. But these assumptions represent both caution flags for the practitioner and hot spots for the researcher. One notorious spot is the adopted model of independent private value model with a commonly known number of agents. In particular subjects' valuations should be further investigated. The present thesis assumed values to be uniformly distributed – but since neither real-world distribution of values is known nor the effect of the distributional assumption, other distributions ought to be studied as well. Moreover, subjects knew the maximum possible value and used this knowledge to create an intuitive ad-hoc strategy. In practice, such a maximum value will not be commonly known. Without such information, the influence of value and hammer price on bids is likely get stronger. The hammer price in turn represents another opportunity for subsequent thinking. In the present study, one market price was considered, which was the same for all orders and which was commonly known. This assumption can be softened in several ways. Orders from A to B, orders from B to A, and package prices could have different but commonly known market prices. It is likely that subjects will still make their bid depend on their value and the relevant market price. On the other hand, a range could be introduced from which the price(s) is (are) drawn according to some random distribution. The number of bidders represent an important issue as well. How important is information on this number? An experiment where the number of subjects is not commonly known but where all other conditions are identical should be conducted. It would deliver a hint whether such information enhances transaction efficiency or not. Last, but not least, of cause, Nash equilibria for Dynamic Alliance auctions with an arbitrary number of bidders and value distribution have still to be discovered.

A

Proofs

Proposition A.1. *DV, TB, TB', and HP are supposed to hold and to be common knowledge to shippers. Suppose all shippers know the hammer price H. Then B_h forms a symmetric Nash equilibrium.*

Proof. Paul assumes that all shippers in $\mathcal{S}$ and $\mathcal{S}'$ stick to B_h. Suppose Paul's value is x. If Paul also sticks to B_h, he makes zero profit if $x < h$ and a profit of $x - h \geq 0$ otherwise.

Case I: $x < h$. Paul cannot gain by deviating from B_h. With a bid $b \geq h$, the probability is strictly positive that Paul will be matched with a shipper who bids h. In this case, Paul's payoff will be $x - h < 0$, so that the expected payoff will be also negative. This is worse as before. On the other hand, Paul cannot gain by bidding b such that $b \leq x$ or $x \leq b < h$.

Case II: $x \geq h$. Paul cannot gain from bidding $b \geq h$ either. If he bids $b > h$, he will still be matched to a shipper in $\mathcal{S}'$ who bid h, but his payoff will be

$$x - b/(b+h) \cdot H < x - h.$$

Hence such bid makes Paul worse off. Paul can also not gain by submitting a bid $b < h$. If he submits such a bid b, he will be matched with probability $1/2$ either to S'_b or to S'_{b+1}. If $b < h - 1$ holds his payoff is zero since in both cases the order package containing his order will not be placed. If $b = h - 1$, his expected payoff is

$$1/2 \cdot 0 + 1/2 \cdot (x - h) < x - h. \qquad \text{(A.1)}$$

Again, deviating from B_h makes Paul worse off. □

Proposition A.2. *The assumptions DV, TB, TB', and HP are supposed to hold and to be common knowledge to shippers. Suppose all shippers know* *the hammer price H. Then no Nash equilibrium exists that does not satisfy (h-SYM).*

Proof. The proof comprises two steps. The first step is to show that there is no asymmetric Nash equilibrium (B, B') with the following property (P).

(P) There is an $x \in \{0, \ldots, M\}$ such that $B(x) \leq h - 2$ and the order of S_x belongs to a package with a reservation price $r \geq H$.

This implies that the only possible Nash equilibria dissatisfying h-SYM could be formed by strategies (B_-, B_+) or (B_+, B_-) that satisfy the following conditions:

$$B_+(x) \begin{cases} < h+1 & x < h+1 \\ = h+1 & \text{otherwise.} \end{cases} \tag{A.2}$$

$$B_-(x) \begin{cases} < h-1 & x < h-1 \\ = h-1 & \text{otherwise.} \end{cases} \tag{A.3}$$

The second step is to show that (B_-, B_+) and (B_+, B_-) are no equilibria. Putting the two steps together, the proof is complete.

Step 1. Assume, for contradiction, that there is a Nash equilibrium (B, B') that satisfies property (P) but not h-SYM. Pick out the according $x \in (0, M)$. Now, some cases must be worked through. In each of the cases, one bidder can be found who can gain from deviating from B or B' if he assumes that all other bidders stick to (B, B').

It is clear that S_x cannot have a negative equilibrium payoff. Otherwise (B, B') would have ceased to be an equilibrium immediately because S_x would then be better off bidding his value. According to the assumptions S_x is matched to a shipper $\tilde{S}' \in \mathcal{S}'$ who submitted a bid $b' \geq h + 2$. Denote the value of $\tilde{S}'$ by $v(\tilde{S}')$.

Case 1: $v(\tilde{S}') < h + 2$. If all other bidders stick to (B, B'), bidder $\tilde{S}'$ is better of bidding $v(\tilde{S}')$, which grants him a non-negative payoff. Hence (B, B') is no Nash equilibrium, in contradiction to what has been assumed.

Case 2: $v(\tilde{S}') \geq h + 2$. Consider S_{h-1}, i. e. the shipper in $\mathcal{S}$ with value $h - 1$. Denote his expected equilibrium payoff by $\pi(B, B', h-1)$. Note that $\pi(B, B', h-1) > 0$ must hold because otherwise S_{h-1} could improve his payoff by bidding $h - 1$. This means, with strictly positive probability, S_{h-1} has a strictly positive payoff from a matching with a shipper S'. Because $r \geq H$ was assumed to hold, shipper S' submitted a bid $b \geq h + 1$. Of course, the payoff of S' must be non-negative – otherwise, S' could gain by deviating from B'. This implies that S' has a value of at least $h + 1$. Consequently, there is one shipper S_z in $\mathcal{S}$ who has a value $z \geq h$ whose order is matched with the order of a shipper S'_y with value $y < z$. Now there are two possibilities, and both lead to contradictions of the assumptions. The first possibility is that S_z has no strictly positive expected payoff. This would contradict the assumption that (B, B') is a Nash equilibrium – S_z could gain by setting his limit equal to $h - 2$. The second is that S_z has a positive payoff, which in turn means that S'_y bids more than $h + 2$. This gives him a loss he could avoid by bidding e.g. zero or his value. □

Step 2. Let B_+ and B_- satisfy conditions (A.2) and (A.3). The following will show that (B_+, B_-) and (B_-, B_+) represent no equilibria because one of two shippers with the highest value M can gain from deviation.

Consider S_M. Let S_M assume that all shippers in $\mathcal{S}$ stick to B_- and all shippers in $\mathcal{S}'$ rely on B_+. Without loss of generality, assume that H is an even number.[1] Note that there are $M - h$ shippers in $\mathcal{S}$ who have a value of at least $h - 1$ and $M - h - 1$ shippers in $\mathcal{S}'$ who have a value of at least $h + 1$. These numbers can be used to calculate S_M's probability to get matched with a shipper in $\mathcal{S}'$ who bids $h + 1$. This probability is

$$\frac{M - h - 1}{M - h}$$

and S'_M's according payoff is $M - h + 1$. Since Paul gets zero payoff in case he is matched to any shipper in $\mathcal{S}'$ with a lower value, his expected payoff from bidding $h - 1$, which is what B_- prescribes, will be

$$\frac{M - h - 1}{M - h} \cdot (M - h + 1) \tag{A.4}$$

But Paul can do better by bidding h. Then, the probability is one that he gets matched with some shipper in $\mathcal{S}'$ who has bid $h + 1$. His according payoff is

$$\begin{aligned} M - H \cdot \frac{h}{H+1} > M - h > M - h - \frac{1}{M-h} \\ = \frac{(M-h)^2 - 1}{M - h} \qquad \text{(A.5)} \\ = \frac{M - h - 1}{M - h} \cdot (M - h + 1). \qquad \text{(A.6)} \end{aligned}$$

Note that (A.6) follows from (A.5) due to the binomial theorem. Consequently, (B_-, B_+) is no equilibrium. Analogously, neither is (B_+, B_-). □

[1] The proof holds just the same for odd numbers or any real numbers if using $[h] := \min\{n \in \mathbb{N} | n \leq h, n \geq m \forall m \in \mathbb{N} \text{ such that } m \leq h\}$.

B

Formulas

B.1 Expected Payoff

Equation (7.15) can be derived from equation (7.10) by introducing all terms in 1.) – 4.).

1. It was safe that 6 subjects were in the market, hence $q_6 = 1$. The probability that exactly $s \leq 6$ bidders are in $\mathcal{S}$ is

$$q_{N,s} = \binom{N}{s} \cdot \frac{1}{2^N} = \binom{6}{s} \cdot \frac{1}{64}.$$

2. Furthermore, $H_j = 50$ holds for all $j = 1, \ldots, N$.
3. The distribution function is $F = U([0, 100])$ with the cdf $f = 1/101 * 1_{[0,100]}$. The probability that Paul submits the j-th highest bid equals

$$p_j := \binom{s}{j-1} \cdot F(B^{-1}(v))^{s-j} * [1 - F(B^{-1}(v))]^{j-1}. \tag{B.1}$$

4. The density function of the j-th order statistic is given by $f_{(j)} :=$ for all $j = 1, \ldots, N$.

C

Experiment

C.1 Translated Instructions

Rules of the Experiment

You are taking part in an economical experiment. The individual payment depends on your result and will be made privately after the experiment. The data from this experiment will be used for behavioral research. It is absolutely necessary that everyone who takes part in this experiment makes his decisions on his own. Every kind of communication among participants is forbidden during the entire experiment and will result in an immediate abruption. If the experiment is abrupted, no regular payment will be guaranteed.

At the beginning of the experiment a computer will assign you to one of two experimental markets. In each of these markets the same 6 market participants will meet in 100 rounds. The identity of the participants will be kept secret.

The experiment is about purchasing fictitious items. The profit from the purchase of an item depends on the value that a participant has in the respective round and the price that he/she pays.

The procedure for each round is the same:

1. Each of the six market participants will be randomly assigned (by a computer) to one of two groups. There is *your group* and the *partners' group*. Your group consists of 1 to 6 market participants and, consequently, the partners' group consists of 5 to 0 market participants!
2. For each market participant a computer draws a value for the fictitious item randomly from all integers between 0 and 100. Each integer between 0 and 100 occurs with equal probability, and the respective values of the participants are independent from each other. Your value tells you how much money you get for the purchase of an item.

3. Every market participant nows his individual value for the item and submits a bid between 0 and 100. All bids, values and corresponding profits are expressed in the Experimental Currency Unit (ECU).
4. A computer arranges the bids of your group by declining bids. The computer also arranges the bids of the partners' group in the same way, provided that there are market participants in the partners' group. In case there are identical bids in the same group, the computer decides randomly.
5. As far as possible, the computer forms pairs from the highest bids in your group and the highest bids in the partners' group. The participant who has submitted the highest bid in your group and the participant who submitted the highest bid in the partners' group will be partners. Accordingly, the participants with the second highest (and third highest) bids will be partners. Partner share the costs for the fictitious item.
6. The costs for a fictitious good is 50 ECU. You get a fictitious item if the sum of your bid and the bid of your partner equals at least 50 ECU. The price you have to pay for the item calculates according to

$$\text{price} = \frac{\text{your bid}}{\text{your bid } + \text{ your partner's bid}} \cdot 50\text{ECU} \tag{C.1}$$

   Note: Your price will always be between 0 and 50 ECU. Prices will be rounded to the next integer ECU.
   If you do not have a partner or if your partner's bid equals 0, you get a fictitious item if your bid is at least 50 ECU. In this case your price will be 50 ECU.

Attention: +The higher your bid is,

7. the higher is the probability that you get a partner.
8. the higher ist the probability that your partner has submitted a high bid
9. the higher is your price at any given bid of your partner.
   +The higher the bid of your partner is, the lower is your price (given your bid).
10. Your round payoff is the difference between your value and your price. If you have not purchased an item your round profit is 0 ECU. Attention: If your bid exceeds your value a negative payoff is possible.

The payment after the experiment accords to the sum of your round profits. The exchange rate is 250 ECU = 1 Euro. Additionally, you get 5 EURO for coming to the experiment on time. Eventual losses will be accounted for.

Information

T1: During the entire experiment, you will neither be informed about your prices and profits nor on the number of market participants that belong to your group.

T2: After each round you will be informed about your price and your round payoff. You will not be given information on the number of market participants that belong to your group.

T3: During the entire experiment, you will not be informed about your prices and profits. At the beginning of each round, you will be informed how many market participants belong to your group.

T4: After each round you will be informed about your price and your round payoff. At the beginning of each round, you will be informed how many market participants belong to your group.

All pieces of information, bids and values of past rounds will be saved in a history. You can look at the history at any time by choosing it from the menu. Additionally, the menu offers you a calculator.

The following figures summarize the procedure of the experiment.

Price (without partner)

=

50 ECU

(if you purchase a fictitious item)

Price (with partner)

=

$$\frac{\textbf{your bid}}{\textbf{your bid + your partner's bid}} * \textbf{50 ECU}$$

(if you purchase a fictitious item)

Payoff

=

Value – Price

(if you purchase a fictitious item)

Figure C.1. Figure in experiment instructions.

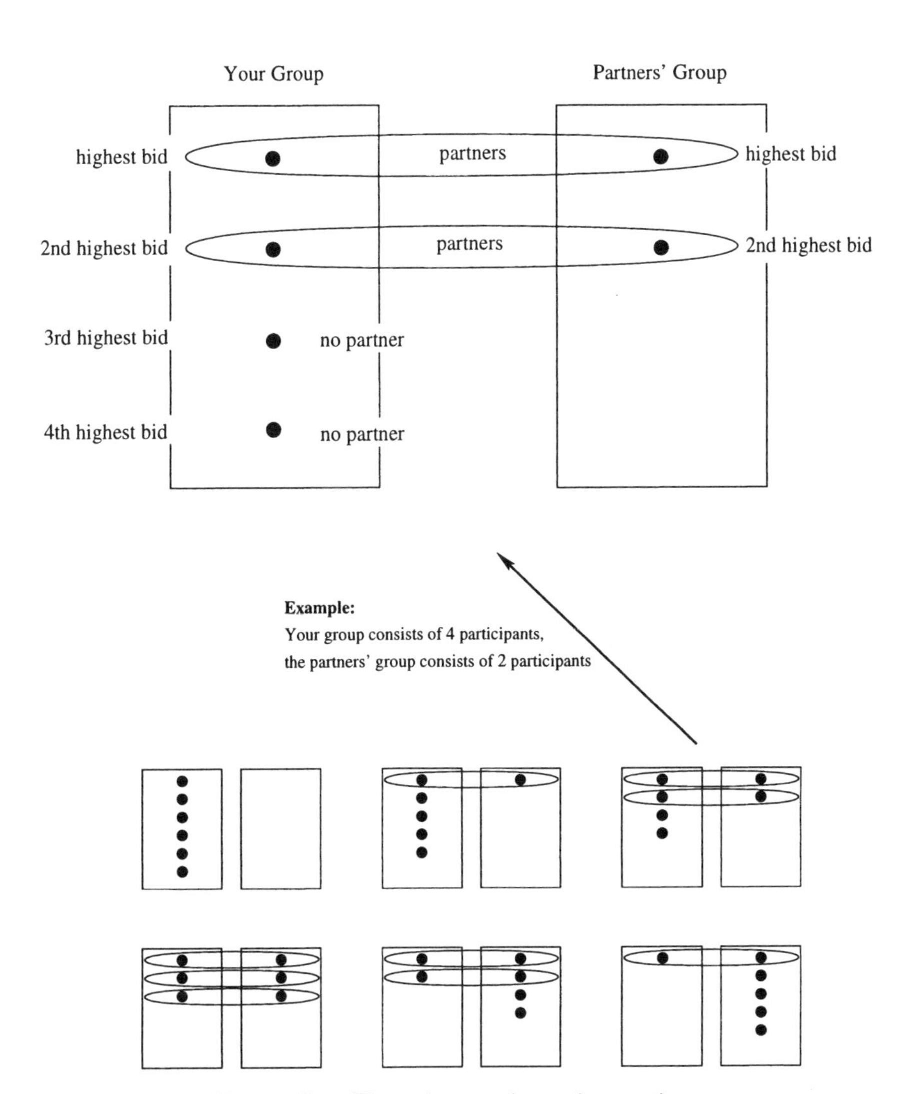

Figure C.2. Figure in experiment instructions.

List of Figures

List of Tables

References

Ashenfelter, O. (1989), 'How auctions work for wine and art', *Journal of Economic Perspectives* **(3) 3**.

Bazerman, M. & Samuelson, W. (1983), The winner's curse: An empirical investigation, *in* R. Teitz, ed., 'Aspiration levels and auctioning for procurement and allocation.', Berlin, pp. 186–200.

Benelog (2001*a*).
URL: *benelog.de*

Benelog (2001*b*).
URL: *benelog.com/app01/jsp/sofunktionierts.jsp*

Binmore, K. (1987), Nash theory i, *in* K. Binmore & P. Dasgupta, eds, 'The Economics of Bargaining', Blackwell, New York, pp. 27–46.

Binmore, K. & Dasgupta, P. (1987), *The Economics of Bargaining*, Blackwell, New York.

Büllingen, F. (1994), 'Einsatz und Diffusion von Telekommunikation im Güterverkehr - Das Beispiel der elektronischen Fracht- und Laderaumbörsen.', Diskussionsbeitrag Nr. 133; Wissenschaftliches Institut für Kommunikationsdienste.

Bretzke, W. R. (2001), 'Die Zukunft elektronischer Transportmarktplätze: Nur wenige kommen durch'.
URL: *mylogistics.net/de/news/themen.jsp?key=news15213*

Bretzke, W. R., Ploenes, P. & Gesatzki, R. (2001), 'Die 'dot.com'-Welle in der Transportindustrie. Eine Studie der KPMG Consulting über Transportmarktplätze.'.

Brewer, P. J. (1999), 'Decentralized computation procurement and computational robustness in a smart market', *Economic Theory* **13**, 41–92.

Bulow, J. & Klemperer, P. (1996), 'Auctions vs. negotiations', *American Economic Review* **86(1)**, 180–194.

Burmeister, B., Ihde, T., Kittsteiner, T., Nikutta, J. & Moldovanu, B. (2002), A practical approach to multi-attribute auctions, *in* 'Proceedings DEXA-2002 Workshop "e-negotiation", IEEE Press'.

Caplice, C. G. (1996), 'An optimization based bidding process: A new framework for shipper-carrier relationships'.

Cassady, R. J. (1967), *Auctions and Auctioneering*, University of California Press, Berkeley and Los Angeles.

DeVries, S. & Vohra, R. (2000), 'Combinatorial auctions: A survey'.

eBay (2001).
URL: *ebay.com*

Economist, T. (2000), 'The container case'.
URL: *economist.com*

esnipe (2001).
URL: *esnipe.com*

Eulox (2001*a*).
URL: *eulox.net*

Eulox (2001*b*).
URL: *eulox.net/cybersped/euloxpublic/leistungen/sprinzipeulox.jsp*

Eulox (2001*c*), 'Terms and conditions'.

Fleetboard (2002).
URL: *fleetboard.de*

Gale, D. & Shapley, L. (1962), 'College admissions and the stability of marriage', *American Mathematical Monthly* **69**, 9–15.

Holler, M. & Illing, G. (1996), *Einführung in die Spieltheorie*, Springer-Verlag, Berlin-Heidelberg-New York.

Ihde, T. (2002), 'Internet-based transportation marketplaces: A critical analysis', *Manuskripte aus den Instituten für Betriebswirtschaftslehre der Universität Kiel* **562**.

Ihde, T., Elendner, T. & Burmeister, B. (to appear), 'How combinatorial auctions work(ed) for daimlerchrysler'.

Ihde, T. & Schild, K. (2002), 'Dynamic Alliance Auctions: Nachfrageaggregation in Internet-basierten Transportplätzen', *Manuskripte aus den Instituten für Betriebswirtschaftslehre der Universität Kiel* **562**.

Isaac, R. M. & Walker, J. M. (1985), 'Information and conspiracy in sealed bid auctions.', *Journal of Economic Behavior and Organization* **6**, 139–159.

Kagel, J. (1995), Auctions: A survey of experimental research, *in* J. H. Kagel & A. Roth, eds, 'Handbook of Experimental Economics', Princeton University Press, Princeton, pp. 501–585.

Kalai, E. & Smorodinsky, M. (1975), 'Other solutions to Nash's bargaining problem', *Econometrica* **43(3)**.

Keynes, J. M. (1923), *A Tract on Monetary Reform*, Macmillan & Co, London.

Klemperer, P. (1999), 'Auction theory: A guide to the literature', *Journal of Economic Surveys* **13(3)**, 227–286.

Klemperer, P. (2002), 'What really matters in auction design', *Journal of Economic Perspectives* **16(1)**, 169–190.

Ledyard, J. O., Olson, M., Porter, D. & Torma, J. A. S. D. P. (to appear), 'The First Use of a Combined Value Auction for Transportation Services', *Interfaces* .

LSXS (2001).
URL: *lsxs.com*

Lucking-Reiley, D. (2000), 'Auctions on the internet: What's being auctioned, and how?', *Journal of Industrial Economics* **48 (3)**, 227–252.

Maskin, E. S. & G., J. (2000), 'Asymmetric auctions', *Review of Economic Studies* **67**, 413–438.

Matthews, S. (1995), 'A technical primer on auction theory i: Independent private values', *Northwestern University. Discussion Paper* **1096**.

McAfee, R. & McMillan, J. (1992), 'Bidding rings', *American Economic Review* **82**, 579–599.

McAfee, R. P. & McMillan, J. (1987), 'Auctions and bidding', *Journal of Economic Literature* **XXV**.

McMillan, J. (1994), 'Selling spectrum rights', *Journal for Economic Perspectives* **8(3)**, 145–162.

Meier, V. S. (2001), 'Anforderungen an Logistikmarktplätze'.
URL: *mylogistics.net/de/news/themen/key/news15779/jsp*

Merkel, H. & Kromer, S. (2001), 'Wie sehen die Frachtbörsen der Zukunft aus?'.
URL: *mylogistics.net/de/news/themen/key/news16610/jsp*

Merz, M. (2002), *E-Commerce und E-Business – Marktmodelle, Anwendungen und Technologien*, dpunkt-Verlag.

Milgrom, P. (1985), The economics of competitive bidding: A selective survey., *in* L. Hurwicz, D. Schmeidler & H. Sonnenschein, eds, 'Social goals and social organization.', Camebridg University Press, Cambridge.

Milgrom, P. (1987), Auction theory, *in* T. Bewley, ed., 'Advances in economic theory: Fith World Congress', Cambridge University Press, Cambridge.

Milgrom, P. (1989), 'Auctions and bidding: A primer', *Journal of Economic Perspectives* **3**, 3–22.

Milgrom, P. & Weber, R. (1982), 'A theory of auctions and competitive bidding', *Econometrica* **50**, 1089–1122.

Mongell, S. & Roth, A. (1991), 'Sorority rush as a two-sided matching mechanism,', *American Economic Review* **81**, 441–464.

Myerson, R. B. (1981), 'Optimal auction design', *Mathematics of Operations Research* **6**, 58–73.

Nash, J. (1950), 'The bargaining problem', *Econometrica* **28**, 155–162.

Neugebauer, T. & Selten, R. (under revision), 'Individual behavior of first-price sealed-bid auctions: The importance of information feedback in experimental markets'.

Osborne, M. & Rubinstein, A. (1990), *Bargaining and Markets*, Academic Press, San Diego.

Owen, G. (1995), *Game Theory*, Academic Press, San Diego.

Polzin, D. (1998), 'Entwicklungspotentiale elektronisch unterstützter Dienstleistungsmärkte im Güterverkehr. Theoretische Grundlagen und Ergebnisse einer empirischen Untersuchung.'.

Rassenti, S., Smith, V. & Bulfin, R. (1982), 'A combinatorial auction mechanism for airport time slot allocation', *The Bell Journal of Economics* **13**, 402–417.

Riley, J. G. & Samuelson, W. F. (1981), 'Optimal auctions', *American Economic Review* **71**, 381–392.

Robinson, M. (1985), 'Collusion and the choice of auction.', *Rand Journal of Economics* **16**, 141–145.

Roth, A. (1979), *Axiomatic Models of Bargaining*, Springer-Verlag, Berlin-Heidelberg-New York.

Roth, A. (1984), 'The evolution of the labor market for medical interns and residents: a case study in game theory', *Journal of Politcal Economy* **92**, 191–209.

Roth, A. (2002), *Al Roth's game theory and experimental economics page.*
URL: *economics.harvard.edu/~aroth/alroth.html*

Roth, A. & Ockenfels, A. (2002), 'Last-minute bidding and the rules for ending second-price auctions: Evidence from ebay and amazon auctions on the internet', *American Economic Review* **92(4)**.

Roth, A. & Sotomayor, M. (1990), *Two-Sided Matching: A Study in Game-Theoretic Modeling and Analysis*, Cambridge University Press.

Rothkopf, M. H. & Harstad, R. M. (1994), 'Modeling competitive bidding: A critical essay', *Management Science* **40 (3)**, 364–384.

Rubinstein, A. (1982), 'Perfect equilibrium in a bargaining model.', *Econometrica* **50**, 97–109.

Sandholm, T. W. (1999), Distributed rational decision making, *in* G. Weiss, ed., 'Multiagent Systems: A Modern Approach to Distributed Artificial Intelligence', MIT Press, Cambridge.

Schmid, B. (1993), 'Elektronische Märkte'.
URL: *businessmedia.org/netacademy/publications.nsf/all_pk/551*

Schmid, B. (1998), Elektronische Märkte – Merkmale, Organisation und Potentiale, *in* A. Hermanns & M. Sauter, eds, 'Management-Handbuch Electronic Commerce', Franz Vahlen Verlag, München.

Schmid, B. (2000), Elektronische Märkte, *in* R. Weiber, ed., 'Handbuch Electronic Business', Gabler Verlag, Wiesbaden.

Schneider, S., Kopfer, H. & Bierwirth, C. (2000), 'Fracht- und Laderaumbörsen im Internet: Von der Pinnwand zum Auktionshaus', *Logistik heute* **(April)**, 2–35.

Schneider, S., Kopfer, H. & Bierwirth, C. (2002), 'Elektronische Transportmärkte - Aufgaben, Entwicklungsstand und Gestaltungsoptionen', *Wirtschaftsinformatik* **(44/4)**, 335–344.

Segev, A., Gebauer, J. & Färber, F. (1999), 'Internet-based electronic markets'.
URL: *citeseer.nj.nec.com/article/segev99internetbased.html*

Shapiro, C. & Varian, H. (1999), *Information rules: a strategic guide to the network economy*, Harvard Business School Press.

Shubik, M. (1983), Auctions, bidding, and markets: An historical sketch., *in* R. Engelbrecht-Wiggans, M. Shubik & J. Stark, eds, 'Auctions, Bidding, and Contracting', pp. 33–52.

Siegel, S. (1997), *Nichtparametrische statistische Methoden*, Klotz, Eschborn.

Teleroute (2001).
URL: *teleroute.de*

van Hoesel, S. & Müller, R. (forthcoming), 'Optimization in electronic markets: Examples in combinatorial auctions', *Netnomics* .

Varian, H. (Dec 12th, 2000), *The New York Times* p. 6.

Vickrey, W. (1961), 'Counterspeculation, auctions and competition sealed tenders', *Journal of Finance* **16**, 8–37.

Wilson, R. (1987), Bidding, *in* J. Eatwell, M. Milgate & P. Newman, eds, 'The new Palgrave: A Dictionary of Economics', The Macmillan Press Ltd., London.

Wilson, R. (1992), Strategic analysis of auctions., *in* R. J. Aumann & S. Hart, eds, 'Handbook of Game Theory, Volume 1', Elsevier Science Publishers B.V.

Wolfstetter, E. (1996), 'Auctions: An introduction.', *Journal of Economics* **10**, 367–420.

Wolfstetter, E. (1998), Auktionen und Ausschreibungen - Bedeutung und Grenzen des "linkage"- Prinzips, *in* Tietzel, ed., 'Ökonomische Theorie der Rationierung', Vahlen Verlag, pp. 139–161.

Wolfstetter, E. (1999), *Topics in Microeconomics: Auctions, Incentives and Industrial Organization*, Cambridge University Press, Cambridge.

Druck und Bindung: Strauss Offsetdruck GmbH